To my dear friends,
Val and George

Peg
Margaret Coleman

Mama Wears Two Aprons
Women in Farming and Farm Marketing

Margaret Marshall Coleman

PublishingWorks, Inc.
2008

Copyright © 2008. Margaret Marshall Coleman. All rights reserved.

All rights reserved. No part of this book may be reproduced or transmitted in any form or by any means, electronic or mechanical, including photocopying, recording, or by an information storage and retrieval system—except by a reviewer who may quote brief passages in a review to be printed in a magazine or newspaper—without permission in writing from the publisher.

PublishingWorks, Inc.
60 Winter Street
Exeter, NH 03833
603-778-9883

For sales and ordering:
Revolution Booksellers
1-800-738-6603 or 603-772-7200

Cover photo courtesy of Brook Photographer and Elizabeth Davis. Back cover courtesy of Neal and Marian Potter.

LCCN: 2007934597

ISBN: 978-1-933002-55-2
ISBN-10: 1-933002-55-7

Mama Wears Two Aprons

Table of Contents

Dedication	iv
Preface	v
Acknowledgments	vii
A Day in The Life of a Farm Wife	1
Prologue: Early Times	3
Chapter 1. Feed Sacks and Farm Women	23
Chapter 2. Upheaval	37
Chapter 3. The Montgomery County Farm Women's Cooperative Market Inc.	43
Chapter 4. Fame and Fortune	53
Chapter 5. Choices	73
Chapter 6. Changes: The 1950s to 1970s	77
Chapter 7. Farm Women of the Twenty-first Century: Making Money	93
Epilogue	101
Appendix A: DNA...And the Hum Heard 'Round the World	107
Appendix B: Government and Agriculture	115
Sources	119
Conversations and Interviews	122
Afterword	123
Index	124

To the King Barn Dairy MOOseum

Preface

All my life I have wanted to be a farmer. In 1980, my husband Jim and I bought 30 acres of abandoned farmland and a broken-down log cabin in Montgomery County, Maryland. I was a farmer at last! The first thing I did was clear a few acres for pasture. Roger Hungerford came to cut the mature trees for his lumberyard. He offered to bring his bulldozer and clean up quickly. But the ripples and valleys, the stone piles hinting of previous inhabitants, and the ruined carriage shed were dear to me. I did not want any of the landscape changed. So I chainsawed selectively, gently clearing the way to three shady, green pastures.

I bought sheep. They multiplied and soon my flock numbered nearly a hundred. But the old-timers said I was not a *real* farmer. I took the ram lambs to market, dyed and spun wool, held dye workshops, invited schoolchildren for field trips, planted a large garden, and took herbs and flowers to market. I joined the Montgomery County Agricultural Advisory Committee and Board. I sold the development rights to my farm. I restored the ancient log cabin on my property and opened a farm bed-and-breakfast.

But still the old-timers told me I was not a real farmer.

Just what *is* a real farmer?

When Barbara McGraw and Jack Davis of the King Barn Dairy MOOseum asked me to write a book about women in farming and farm marketing, I declined. But then I visited the Farm Women's Market in metropolitan Bethesda,

Maryland, and talked with some of the stall owners. Barbara and her brother, Garner W. Duvall, told me about their late grandmother, who was the first president of the market. The more I learned about Macie King, the more interested in the subject I became. Macie King was a farmer and a gracious lady, and as I studied her life, I discovered more women who were like her—strong, competent, intelligent women. Women who were willing to take chances, make hard choices, and work long hours at arduous labor to save their families and their homes.

These women were real farmers. Although many of their family farms have dissolved into suburbia, the economic value accrued by these courageous women made a lasting difference. Their life stories beg to be told.

Margaret Coleman
Pleasant Springs Farm
Boyds, Maryland

Acknowledgments

This book could not have been written without the cooperation and enthusiasm of the people I interviewed, who gave of their busy lives to talk with me. My grateful appreciation begins with Barbara McGraw and Jack Davis, who convinced me to write the book and shared their family memoirs. Garner W. Duvall, Mrs. McGraw's brother, lent me the invaluable scrapbook of their late grandmother, Macie King.

Still in its embryonic stage, the new enterprise needed a name. Picturing the feed-sack apron worn by Sarah Hilton and the white aprons of the market ladies, Betsy Lyman tossed out, "Mama Wears Two Aprons." Thank you, Betsy.

Many others told me about their lives, and the farming experiences of their mothers, grandmothers, and stepmothers. Each story was a precious gem, a tribute to the wisdom, courage, and strenuous labor of farm women. Thank you, Donovan Cunningham, Elizabeth Daniel Davis, Louise Barnes Duvall, Debbie Hardisty, Delbert Foster, Betty Jean and James Wriley Jacobs, Peggy Johnson, Cecile King Jones, Frances Wilmot Kellerman, Caroline and George Lechlider, Dr. Susan Moxley, Donna Oden, Jean King Phillips, Neal and Marian Potter, Nona Schwartzbeck, Dr. George Sopko, Elena Stamberg, Barbara Riggs Stiles and Stanley Stiles, Janet Stiles, Olive Gladhill Stup, Anne Tankersley Sturm, and Madeleine Wojokowsky.

Special thanks to the staff of Peerless Rockville; the Sandy Spring Museum; Douglas Tregoning of the Montgomery County Extension Service; the Montgomery County Historical Society; the University of Maryland; and the United States Department of Agriculture, for the use of their archives. And to Peggy Carr of Sir Speedy, Rockville, for her computations and photographic duplications. The officers and board members of the King Barn Dairy MOOseum read the manuscript and lent encouragement.

Thank you to Judy Stone of Stone Graphics for her computer wizardry.

Without the financial support of my sponsors, *Mama* would have remained locked in my computer. Thank you to Mr. and Mrs. John McGraw, to Jeremy Criss of the Montgomery County Department of Economic Development, Sugarloaf Regional Trails, and to the King Barn Dairy MOOseum for your generous and essential contributions.

A very special note of gratitude to Dr. Carla Easter, National Institutes of Health, National Human Genome Research Institute, for reading and fact checking of Chapter 8, and to Maggie Bartlett, NIH.

Last but not least, thanks to my husband, Jim Coleman, who read the manuscript and supplied encouragement and thoughtful commentary.

A Day in the Life of a Farm Wife

She wakes early, builds the fire and makes pancakes and sausage for breakfast. She milks the family cows, strains the milk, skims the cream, washes dishes, makes beds, empties chamber pots, does the laundry and hangs it out to dry. Now she makes dinner of pork, potatoes, cabbage, apple pie, and coffee. She washes dishes, finding she first must prime the cistern pump to get water. Dishes done, she fills lamps with oil, washes their chimneys, takes the clothes off the line, feeds the chickens, and gathers eggs. Now she makes supper, washes dishes, makes dough for tomorrow's baking, and helps the children in their studies.

—Anonymous author, 1901; included in "Women in Agriculture," presented by Dr. Vivian D. Wiser, USDA conference, January 1976. Source: Archival files of USDA.

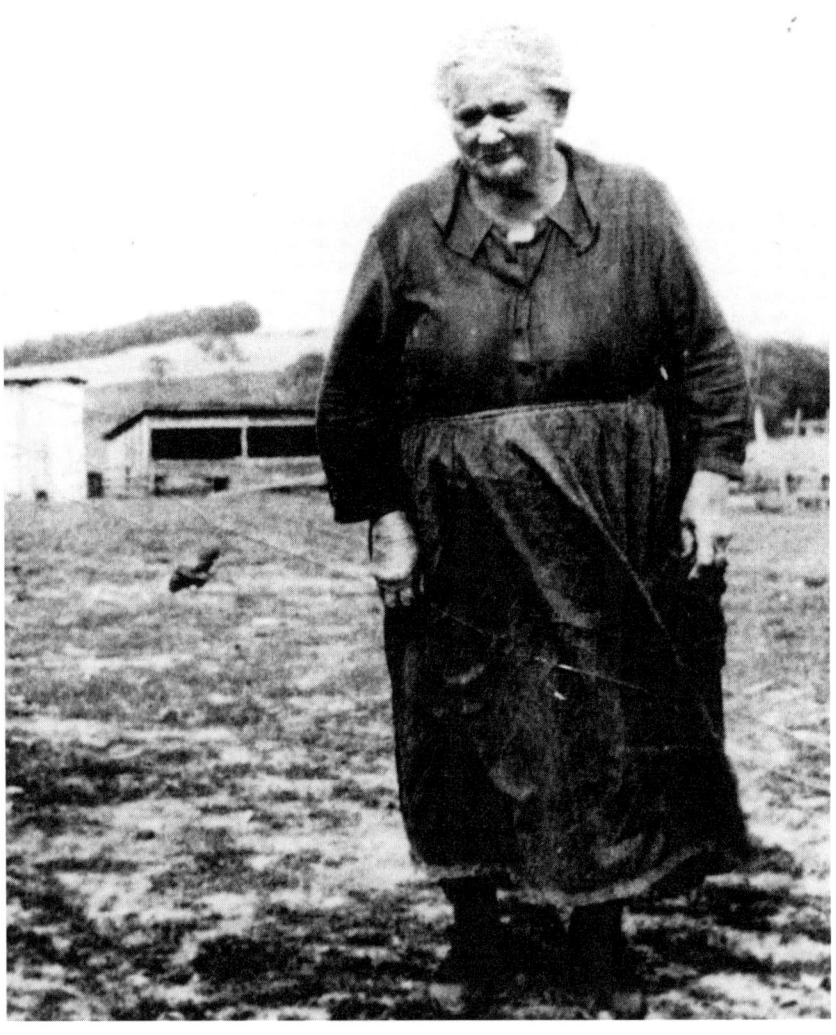

Farm Woman, 1930. Sarah Elizabeth Brown Hilton personifies the farm women in the 1930s. The wind is blowing her feed-sack apron, and her stooped shoulders lean into the breeze. Her hands, swollen from years of hard work, made most of the clothes her family wore; slaughtered and cleaned chickens for dinner; and raised geese and turkeys, too. Sarah maintained a large vegetable garden, canning the produce or storing it in the root cellar. Her apron is tied with string from the feed sack and could become a handy container for eggs, tomatoes, cabbage, or a small child. Courtesy of Jean King Phillips.

Prologue
Early Times

Women were not invented until the 1950s. Or so it seems to the archival reader, searching for accounts of early farm women. Primary sources, however, reveal a different picture. At the heart of every successful farm was a woman.

James Wriley Jacobs put it this way: "Women were the glue that kept things together."

Caroline Lechlider said, "Without (a woman) there would be no farmer. Whether sister, wife or mother, anyone who has done well" has a woman in his home.

Diane Savage Greary said, "Woman's role was the rock on which the silo was built, the cornerstone of each barn, the fuel that fed the engine that made each farm run."

In America, as elsewhere in the world, society was based on John Locke's *Two Treatises of Government*, first published in 1689. Locke explains that men, not women, own property. It is their natural right because of their sex. Locke wrote, "though the earth and all inferior creatures be common to all men, yet every man has a 'property' in his own 'person.' This nobody has any right to but himself. The 'labour' of his body and 'work' of his hands, we may say, are properly his."

But not hers...Women as well as men labor on the land, but women cannot own the property, according to Locke. His philosophy covered native occupants of the land, too.

According to Locke, their lands were in a state of Nature and could be appropriated by those with an advanced political society having a system of labor-based property.

Thus ensconced in Locke's descriptive patterns, a woman marrying a farmer accepted the fact that the farm on which she labored and which she often loved; her name and her children's names; the bed in which she slept—all were her husband's by right.

Montgomery County, Maryland, was settled in the early eighteenth century, and is a prototype of the United States in many ways except two. In the first place, Montgomery's next-door neighbor is Washington, D.C., a powerful force that, during the early years, allowed the intimate exchange of information between the rural population and the nation's political leaders. Later, being next door to hundreds of federal employees with regular paychecks somewhat ameliorated the effect of the Great Depression. In the 1950s, Washington sent highways radiating like beams from a laser to every corner of the country, mowing down farms that were in the way, and sowing the land with houses and commercial structures from the Atlantic to the Pacific.

Second, in the early 1980s, Montgomery County became first in the nation to declare agriculture as the preferred land use on more than 90,000 acres of farms and open space.

Of equal importance to the history and development of Montgomery County agriculture is the Society of Friends in Sandy Spring and Brookeville. Because the Quakers wrote down everything and saved it—and aided by the happy tradition of today's Quakers, who generously share their archival treasures—researchers today enjoy a unique legacy of primary sources.

In 1884, Quaker William Henry Farquhar began to publish a continuing series of volumes, *Annals of Sandy Spring*. From the first, Farquhar speaks of farm women:

"Domestic and agrarian abilities of women were highly regarded and contributed directly to . . . survival of the family unit."

But what exactly did Quaker women do?

From memories and notes of early days, Farquhar comments on farm women in a Quaker society. Settlers found danger all around them, and survival could be ephemeral. Farquhar wrote, "The wolf and the bear ate their young swine and lambs; and the panther made the lonely paths through the woods perilous to them and their children."

Like a fairy tale by Grimm, Farquhar told of yellow eyes glowing in the dark and an axe found the next morning in the skull of a young panther.

Men and women living on farms contributed in different ways. Men slew the dragons and women made the dinner, raised the children, and swept the floors at a time when water came from a spring on the hillside and winter drifted through the cracks in the log cabin, misting the cradle with snow.

Among the earliest Quakers to arrive in the territory were Deborah Snowden Brooke and her husband, James Brooke. Next door to Sandy Spring, they established a Quaker village named Brookeville. With their diligence, the area grew more civilized so that a descendant, Roger Brooke IV, raised a family that included three remarkable daughters, Hannah, Sarah, and Mary. In the late 1700s, they married, respectively, Isaac Briggs, Caleb Bentley, and Thomas Moore, three innovative Quaker men. Thomas Jefferson, speaking of Hannah's husband, said: "Isaac Briggs, in point of science, in astronomy, geometry and mathematics, stands second to no man in the United States."

Hannah and Isaac protected President Madison and the country's gold as the British busily burned the White House. Nearby, Hannah's sister, Sarah Bentley, and Sarah's husband,

Caleb, housed Dolley Madison and as many treasures as the first lady could carry when she fled the redcoats.

With wives to worry about their bodily needs, farm men who were so inclined were freed to focus on inventions. In 1791, Mary's husband, Thomas Moore of Waterford, Virginia, came to Sandy Spring and bought a farm. Faced with poor soil, Thomas added plaster of paris, plowed eight inches deep, and planted nitrogenous cover crops. The next year he plowed eleven inches deep and was rewarded with a remarkably high yield. Being a good Quaker, he wrote down everything, complete with dates and exact yields. For this and other important achievements, he was elected to the Philadelphia Agricultural Society in 1809, the only Montgomery County man to be so honored.

Besides being a progressive farmer, Thomas was an inventor. All early farms had a requisite cow or three, and the Moores were no exception. Every morning, Mary carried the milk from the barn to the stone springhouse, which had a trough on three sides, through which flowed water, "a lovely spot with wild cherry trees and grapevines about a fine cold spring," as she wrote in her diary.

Mary set her pans of milk in the trough. The next morning she skimmed cream off the top, gave the fat-free milk to the pigs, churned butter with the cream, and left the golden chunks in the springhouse to solidify. Thomas Moore collected her butter along with other farm produce, and drove to the Georgetown market on "M" Street, about 15 miles southwest. But in July when he tried to sell Mary's butter, the nice, firm chunk was runny and sour. No one bought it. Mary, who spent precious hours making that butter, was not happy, and neither was Thomas. Out of necessity came an important invention.

Thomas made a square box of tin and an oval box of wood to fit inside the tin box. After filling the empty corner spaces with ice from the icehouse, Thomas placed

Mary's butter in the oval box. Over the top of both boxes, Thomas laid a rabbit skin, heavy with thick fur. Over the rabbit skin, he placed burlap. He secured the box in the wagon bed between his cabbages and rutabagas, hitched up the horse, and off to market he went. Every one of those solid chunks of sweet cream butter sold, and with each purchase, Thomas included a free piece of ice.

Thomas obtained a patent and is credited with inventing the first portable refrigerator.

Creating a home in the late eighteenth century was an enormous achievement. Through awkward times of pregnancy and childbirth, illness and death, washing clothes and linens by hand, keeping wood stoves stoked, preparing endless meals, and washing unfathomable numbers of dishes, Mary kept Thomas happy and content, able to focus on matters outside the house. He was the inventor, but she was the "glue" that kept the home together and made it possible for him to focus on invention and plowing those fields.

The three brothers-in-law, Moore, Briggs, and Bentley, founded a mill town, Triadelphia, and harnessed energy from the Patuxent River to grind wheat and saw logs.

As it was their tradition to learn from others by sharing knowledge, the Sandy Spring Quaker men founded the Sandy Spring Farmers Society (1799), the American Board of Agriculture (1803), the Montgomery County Agricultural Society (1822), the Sandy Spring Farmers Club, the Enterprise Farmers Club (1867), and the Montgomery Farmers Club (1873).

Sarah Briggs Stabler, the daughter of Hannah and Isaac Briggs, is the first woman farmer in the Montgomery County annals. Sarah and her husband, James P. Stabler, wrote to France and ordered silkworms, their eggs and cocoons, and the silkworm's food source, white mulberry trees, *Morus alba*. The white mulberry is different from the native *Morus rubra* and does not flourish in this area. Sarah must have carefully nurtured the trees, because there were plenty of leaves for the tiny indoor livestock.

The silkworms fascinated Sarah. She studied them as they happily chowed down on their mulberry leaves. When the larva became fully mature, they spun their cocoons, moving their little heads around continuously for three consecutive days in order to extract the filament-producing substance from a double set of glands, set one on either side of their small bodies. The worms stayed put for ten to twelve days, and then, at the pupal stage, were ready to escape. By moistening one end of the cocoon, each wee pupa could push aside the fibers and make an opening. Out slid a perfect moth, *Bombyx mori*, white in color, about a half inch long. It puffed out its tiny wings to dry, and flew around a bit. Eggs were laid and fertilized and the individual *Bombyx mori* died, cycle completed.

Sarah and James harvested the empty worm casings. To melt the chitin holding the filaments together, Sarah heated the cocoons by putting them in the oven of her woodstove "when you take out the bread."

Already accomplished at spinning wool, Sarah managed the intricate task of opening the cocoons to reveal the 800 to 1,200 yards of fine fiber, degumming and spinning the delicate substance into a strong, lustrous yarn. Next came dyeing, followed by weaving or knitting the yarns into fabric. Silk takes dye exquisitely, and Sarah must have loved to watch the dull, natural silk take on brilliant tones and shades, blending into an outrageous kaleidoscope in her dye pot. Yellows, gold, a hundred shades of blue and green, scarlet to pale pink—tints and hues appearing like magic charmed her eye.

Indigo had been used for thousands of years by this time, and it was a woman farmer who grew the first crop in the new nation. Eliza Lucas Pinckney of Wapoo Creek, South

Carolina, planted seeds of *Indigo tinctoria* on her 600-acre family farm. Through trial and error she learned to process the plant into the rock-hard stage that creates lust in the hearts of dyers. Eliza distributed seeds to neighboring planters and, by 1770, over a million pounds of indigo was shipped every year from South Carolina to England. But the dye plant harvest faded during the American Revolution, giving rise to cotton as the favored crop. Indigo production was abandoned.

Most of Sarah's dye came from her garden—weld (dyer's rocket) for blue; alkanet, borage, and pokeweed for purple; madder and lady's bedstraw for red; marigolds, genistra, and goldenrods for yellows. Dipping blue-dyed silk into the yellows gave a variety of greens.

In 1836, Sarah and James listed two firms, the Montgomery County Silk Company and the Federal Silk Company. But when James died in 1840, Sarah found the heavy, monotonous work was too much for her to do alone. It was difficult to keep the white mulberry trees and the silkworms alive, and she could not produce enough silk to make the business pay. The Montgomery County silk industry disbanded.

In 1857, the farm women of Sandy Spring formed an organization that exists to this day with few changes. Mary L. Roberts conceived the idea, inviting eleven of her neighbors to her home, Sherwood, to introduce her idea. True to Quaker tradition, Mary wrote down everything.

"Desirous of promoting their own and others' improvement," she wrote, the group would be called the Women's Mutual Improvement Association. The society's purpose was to share information "to elevate the mind, increase happiness, lighten the labor or add to the comfort of one another, or our families or neighbors."

There was to be no agenda, no regular program nor chosen subject. Each member was expected to come to the meeting prepared with something to contribute to the collective knowledge of the Association. There would be just one elected officer and that was, of course, the secretary, the person who wrote down everything. At the following gathering, the secretary would read the minutes of the preceding meeting, then call first upon the hostess, who was expected to give a "pithy sentiment." Then the officer would name each woman in turn and write all of the comments in a well-bound journal.

From that day forth, the record of each monthly meeting of the Association has been headed by the name of the place and the number of the event, now in the thousands. No addresses were needed; all the ladies knew how to find each other's farms—Bloomfield, Falling Green, Cedar Lawn, Brierly, Springdale, Sharon, Brooke Grove, and Cherry Grove, as well as Mary's house, Sherwood.

To open an Association journal is to reveal daily lives, concerns, joys, and sorrows of mid-nineteenth-century farm women. Clearly they were involved in agriculture and immersed in sharing information and experiences.

Sally Pierce advised on controlling curculio worms in plums: Dig a trench around the tree in autumn and remove all the dirt to expose the roots; next spring, fill in the space with fresh dirt.

Lydia Thomas told how to transplant roses.

Deborah Lea told how to make good soap: Save all fat scraps, cover them with strong lye, add more lye from time to time, and stir occasionally "til soap comes."

Contributions could be in the form of thoughtful advice. Mary Roberts—whose predilection to underlining was prodigious—wrote that women must "employ well such <u>talents as we have</u> without being discouraged because they are not greater." Later she added, "To try to be happy is the way to become so, and is one of our <u>duties</u>."

Printed material was welcome, such as a newspaper clipping discovered in "grandfather's journal." Recipes, the report of a trip to Niagara Falls, a letter from England, poems—all found their way into the minutes of the Association.

Sadness drifted into Sandy Spring, but to share one's grief was a comforting salve for a broken heart. In 1858, Mary Roberts wrote, "The Angel of Death has visited us since last we met; he has gathered some of the brightest flowers ... and why? They were wanted in Paradise, they were too good for this world."

She read a poem, echoing the ancient and current sorrows of women:

> And the mother gave in tears and pain,
> Those flowers she most did love;
> She knew she should find them all again,
> In the fields of light above.

A few years later, war drums rumbled across the peace-loving farms of the Quakers. Confederates had crossed the Potomac at Edward's Ferry and were coming their way. Rebels were in Rockville! In Mechanicsville! They detained Quaker relatives near Laytonsville until "our soldiers came and chased them out." Rebels were taking the horses, and had burned and destroyed the home of Postmaster General Montgomery Blair.

Habitual writers and highly literate Quaker women wrote their personal life experiences. In a beautifully written memoir, Mary Briggs Brooke told about her life during the years 1868-1890. Water was pumped by hand from a well under the back porch. She was cold in the winter. Trapping rabbits, gathering chestnuts, and eating chickens and beef cattle were all a part of nutritious farm life. Mary Brooke described a hog slaughtering. The day before, logs were laid out "like a log cabin" with soapstones in the center. After a night of baking, the soapstones were very hot. They were picked up and dropped into a hogshead (a barrel large enough to hold 63 to 140 gallons of water) until the water became very hot—hot enough to scald the hair off a hog. The slaughtered pig was lowered into the hogshead head first, then tail first "with difficulty." Pigs' bladders were saved and hung in the cellar to use as ice bags in case of sickness.

Bladders made good toys. "We children were allowed to have some of them and it was great fun to play with them," she wrote.

Mary Briggs Brooke wrote about the good times and the bad ones, too. All the children came down with measles, but her sister Eliza had the worst case of them all. "Eliza had convulsions for twelve hours," Mary wrote. There was not a lot that could be done for her. Bridget, wrote Mary, put Eliza in a tub of mustard and water.

Father was so upset, he "lay down on the foot of the bed crying!"

But Eliza did not die. Unlike many sufferers from measles, Eliza recovered.

Into the journals of the Association again, mentions of farming were subtle but not absent. The ubiquitous curculio worm received several suggestions for extinction. To control bugs in honeysuckle, Rachel Gilpin used tobacco "in some form," either sprinkled or smoked. Attracting frequent mention was the scourge of Japanese beetles.

Toward the end of the century, the shaded lanes and cool brooks of Sandy Spring began looking good to city dwellers sweltering in the summer heat. Hearing the opportunity for financial gain knocking at their doors, the farm women of Sandy Spring began taking in summer boarders. In August 1904, Sallie Chandlee Bentley described the situation charmingly thus:

The Summer Boarder

Oh! the sunny days of summer.
Oh! the pleasant days of summer,
With its welcome fruit and flowers
And its never failing cheer.
We must put our house in order,
For now comes the summer boarder,
For the summer is upon us
And the boarder draweth near.
Soon they'll fill the lawn and hammocks
And the porches to o'erflowing,
They'll embroider, golf, play tennis,
And new novels will be read;
Then the housewife's heart grows lighter,
Gold prospective shines the brighter,
Yet she wears an air abstracted,
For those boarders must be fed!
When she seeks her couch at night-time
With the hope of restful slumber,
She begins to plan her menus
With a method quite intense,
Meats and entrees, desserts sweeter,
Trusting some surplus will greet her,
For she must consider dollars
And she has to think of cents.
．．．．．．．
Here she drifted off to dreamland
With a soft contented sigh,
For she dreamed her guests ecstatic,
With loud praises most emphatic,
Feasted on this whole collection
Piled upon the table high.
Now the moral of this lesson,—
For each thing has some wee moral,—

> If we want our summer boarders
> And enough to make it pay,
> If we really think we need them,
> We must feed them, feed them, feed them!
> Spare not fruit, cream, chickens, butter,
> And the boarders come and stay.

By 1912 many Quaker women were demanding equality at the polls. But not everyone agreed this was the right thing to do. At a meeting of the Association, Estelle Moore worried about the prediction of an unidentified scientist: "Girls of the future will be bald," she read. This dreadful state of affairs was the direct consequence of women becoming "almost equal to men." For baldness "is a sign of intellectual powers and advance of civilization."

Choosing to sacrifice their hair if need be, in 1920 the Sandy Spring ladies joined with women in every one of the United States and voted.

As to farming, it appears that actual hands-on farming and farm marketing was left to the men of Sandy Spring. But to the women fell the task of making their community better. Men might invent an icebox, but the women nurtured their husbands and improved the lives of themselves and society.

Food mattered. After plowing ten inches deep behind a horse all day, Thomas Moore longed for a good dinner. Perhaps Hannah prepared one of these dishes on her wood-fired cookstove.

> **Red Flannel Hash**
> 4 beets, cooked
> 2 potatoes, cooked
> 2 Tablespoons butter
> 1 1/2 pound roast, minced onion, poached eggs, cream

Cook onion in a little butter. Add remaining ingredients except the eggs. Mash together well. Spread in flat baking dish. Melt butter with a little cream. Brush over the hash. Brown in oven. Top with poached egg.

Indian Pudding
2 1/2 cups milk
3 Tablespoons corn meal
1/2 cup molasses
1/2 teaspoon cinnamon
1/4 teaspoon ginger
1/2 teaspoon salt

Scald milk. Add corn meal. Add molasses. Cook 10–15 minutes until thick. Add remaining ingredients. Bake in slow oven for 45 minutes.

(Recipes courtesy of the United States Department of Agriculture, Beltsville, MD, dated 1787.)

While Sandy Spring had its Quakers to develop and pass along knowledge of agriculture and farm life, the rest of the county had the University of Maryland Extension Service.

President Woodrow Wilson signed into effect the Smith-Lever Act of 1914, establishing the Cooperative Extension Service as a partnership of land grant universities and the United States Department of Agriculture. The Act created a powerful tool designed to help individual farmers. Extension Service agents, stated the Act, were to provide "intensive on-the-farm educational assistance in appraising and resolving . . . problems."

Each county was to have one chief extension agent and assistants as needed. Under the Extension Service agent's supervision would be a home demonstration agent, whose job it was to educate rural women in home arts, teaching them safe methods of food preparation and preservation plus every craft from crocheting afghans to installing zippers.

Indeed farm women needed help, although perhaps it was more in the how-to-stretch-the-day department. According to an account written in 1916, a typical farm wife and daughter did all the housework, laundry, and cooking for four hired men, the farmer, and themselves. And they prepared for winter, setting aside or canning 300 quarts of fruit, 100 quarts of vegetables, 4 bushels green beans, 3 bushels onions, 400 head cabbage, 10 bushels turnips, 6 bushels beets, 6 bushels tomatoes, 36 cauliflower, and 7 bushels corn. They used 144 pounds coffee, 50 bushels potatoes, 1,800 pounds pork, 200 poultry, 520 dozen eggs, and 313 pounds butter!

Because they were required to make a written annual report, extension agents left a valuable paper trail. Fred J. Van Hosen was the first extension agent to leave a record in the Montgomery County files. In 1918, agent Van Hosen found the job quite taxing, plagued as he was by war, influenza, and "Farmerettes."

Looking out of the faded photograph is a dour-looking man. Van Hosen complained about the lack of office help and the struggle to raise money by the sale of Liberty Bonds and the War Community Fund. County women, he wrote, were knitting socks for the Red Cross to distribute to soldiers, but more women were needed to help the doughboys in the trenches overseas.

Here at home, wrote Van Hosen, the influenza epidemic "hit the county hard," with more than one thousand cases at one time. During a period of four weeks "everything was at a standstill." Then a second flu wave "of almost epidemic proportions" swept the county.

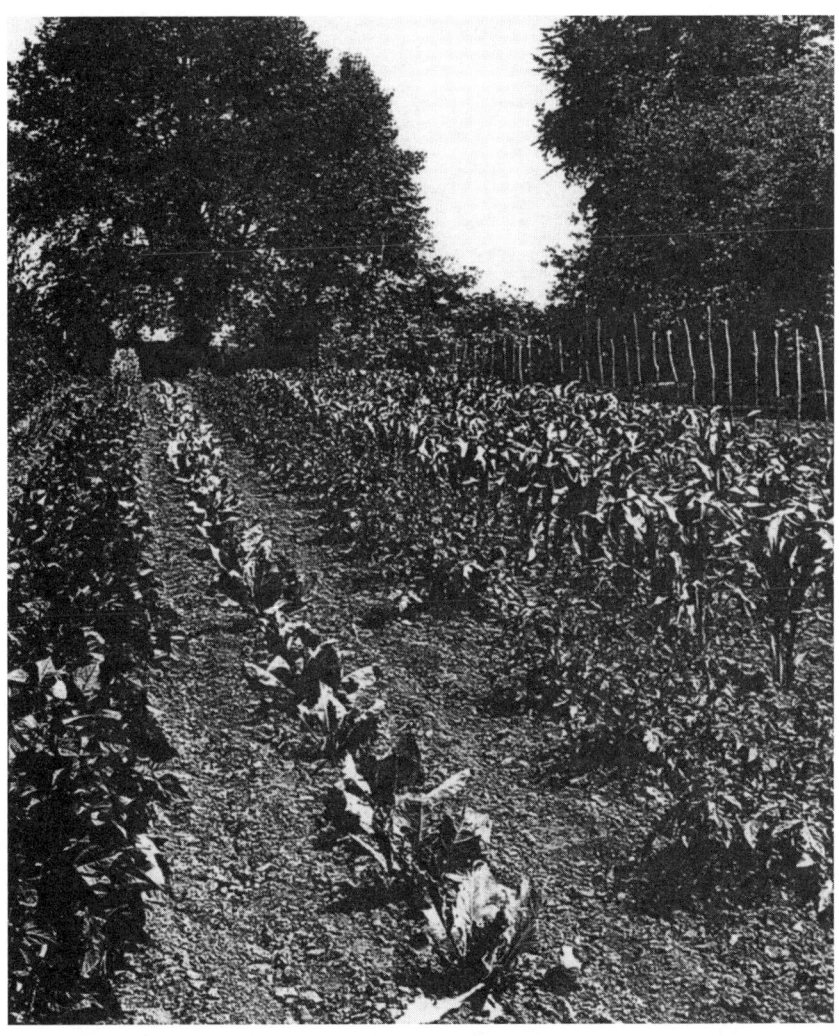

Farm Family Garden, 1922. "No need to fear the food administrator with a garden like this on every farm." No need to go to the grocery store either. Courtesy of Montgomery County Extension Service archives, Derwood, MD.

To make matters worse, "A colony of Farmerettes was established near Rockville but failed to attract much attention except by the novelty of their costumes." Van Hosen wrote that their clothes were funny, but the Farmerettes got a lot done. In these wartimes, wrote the agent, the women packed and graded apples, but most of their time was spent working on a state road.

Agent Van Hosen predicted obesity in the future if residents would not stop eating fried food: "Montgomery County," he wrote, "probably produces more fried chicken per square mile than any other county in the United States."

"Wheat is the principal cash crop," Van Hosen wrote in 1919. "Dairying is one of the large industries of the county," and there were 17 commercial orchards and 8,000 fruit trees.

While men were harvesting wheat, and Farmerettes were packing apples, other farm women were busily spinning. With 2,000 sheep on 90 sheep farms, the wheels made a sweet song promising warm, cozy nights beside a cheerful woodstove.

Extension Service agents were really not needed in Montgomery County, according to Van Hosen. Montgomery "is not a county in which helping the poor farmer works. Our farmers are quite capable of helping themselves."

The next agent had a lot more fun at work. Wardney C. Snarr had a mouth that turned up at the corners, and a twinkle in his eye. His first year, 1922, he counted 2,145 farmers in the county, 35 apple orchards, and 23 peach orchards. And 10 acres of rutabagas! Snarr's reports bubbled with enthusiasm. He said nice things about farmers and especially admired the Quakers.

"The Sandy Spring community is one of the most interesting communities in the United States," he wrote, mentioning the oldest men's club and the oldest women's club in the nation. "Every member attends each meeting," Snarr wrote. Quaker farmers helped the local economy. Their Cooperative Marketing Association of 36 men sold $21,629.34 worth of milk during one quarter—the equivalent of $265,603.15 in 2007.

It was Agent Wardney Snarr who wrote the first notes about a female farmer. Elizabeth S. Jones, The Briers, Olney, requested Snarr's advice about pigs. Young Elizabeth, a member of the Pig Club, wrote to Snarr, and he filed the little girl's letter with his 1923 report.

> I do not feel that I got a very good pig in the first place, as she is too short and chunky. The first year she had a litter of five, but only three this year. I have decided to sell my sow for pork and buy another one . . . A new pig costs $75 and I do not know whether it is good business or not to put practically all of my profits in another pig. I wonder if I can get this sow insured?

Snarr did not record his response, but the first picture of a woman in the county extension records, young or old, is of Elizabeth S. Jones with her pig.

And it was Snarr who made the first recorded reference to a woman who would influence generations and change the course of many lives for the better. Miss Blanche I. Corwin, home demonstration agent for this county, according to Agent Snarr, was in charge of the Women's Department at the Poolesville Fair.

In 1932 Snarr was replaced by Otis W. Anderson, a charismatic man of many talents. O. W. Anderson would play a big part in the development of agriculture in Montgomery County. In his 1934 annual report, Anderson

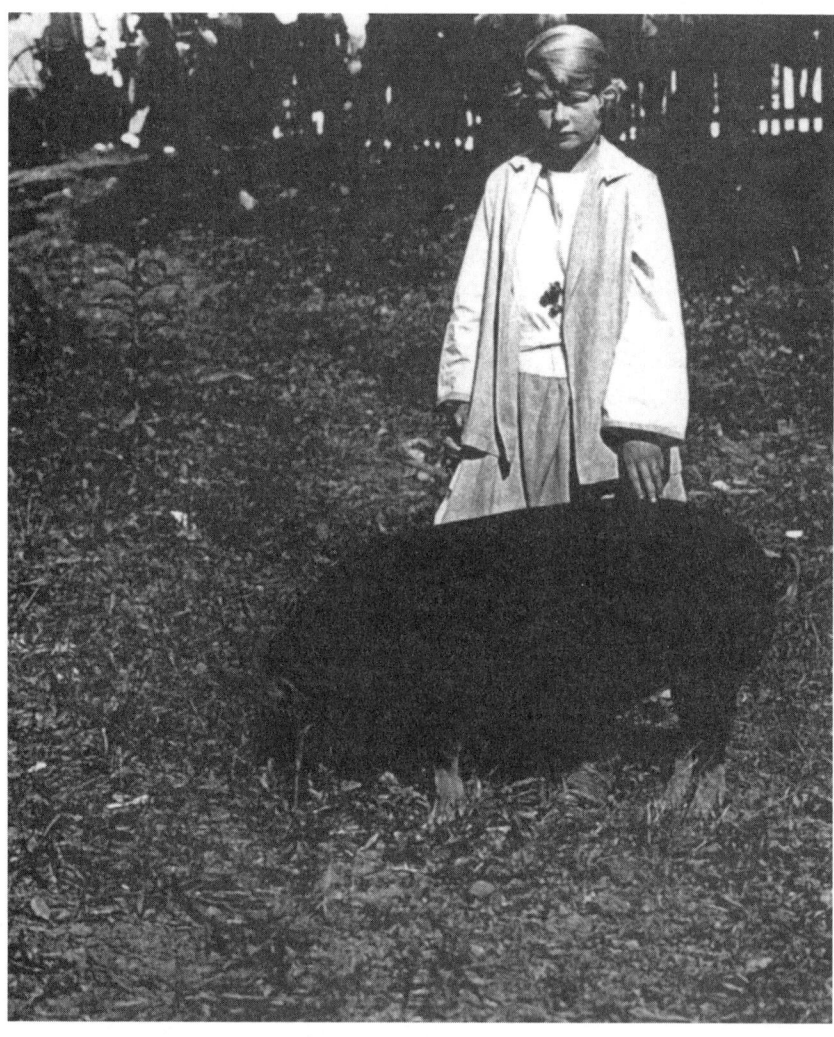
Elizabeth Jones and her pig. Courtesy of the Montgomery County Extension Service archives.

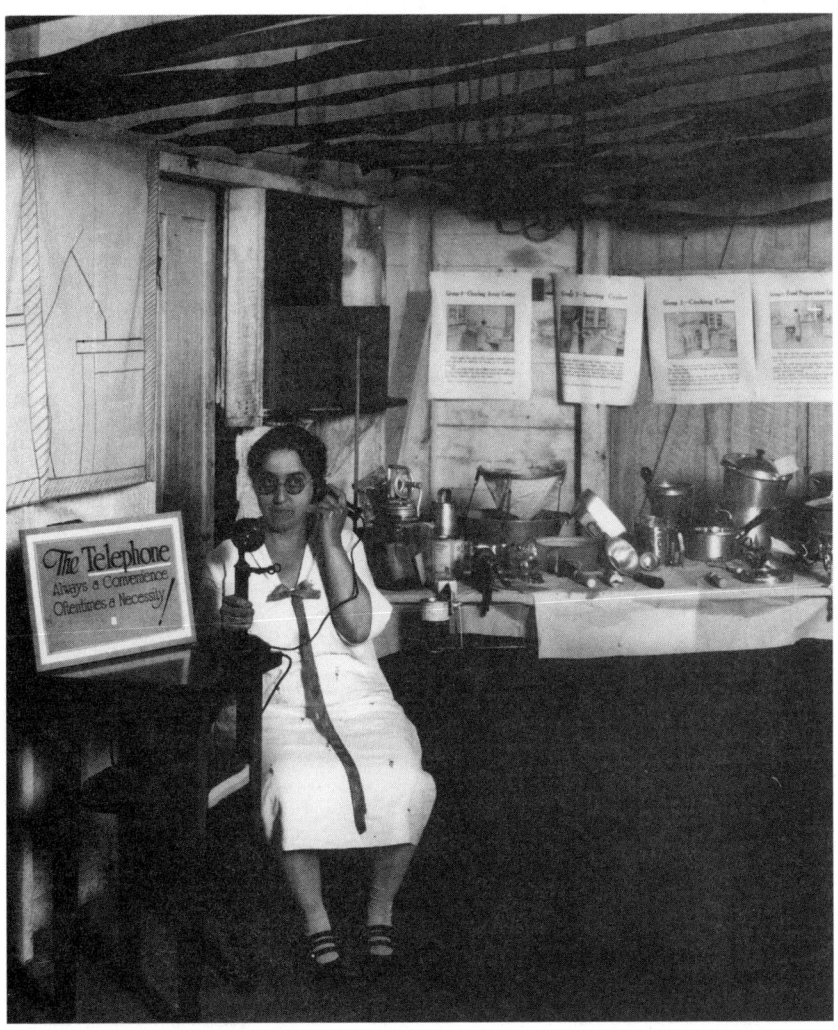

Blanche Corwin, Montgomery County Home Demonstration Agent, 1922–1932, shows how to use the new-fangled telephone. Behind her are the implements she uses to demonstrate modern cooking methods. Courtesy of Peerless Rockville.

wrote that Montgomery County had adopted a milk ordinance giving the power of authority to the Health Department, an act with far-reaching effect.

Committed to helping local farmers, the extension service began to expand, adding talented agents with ideas that were often amazing.

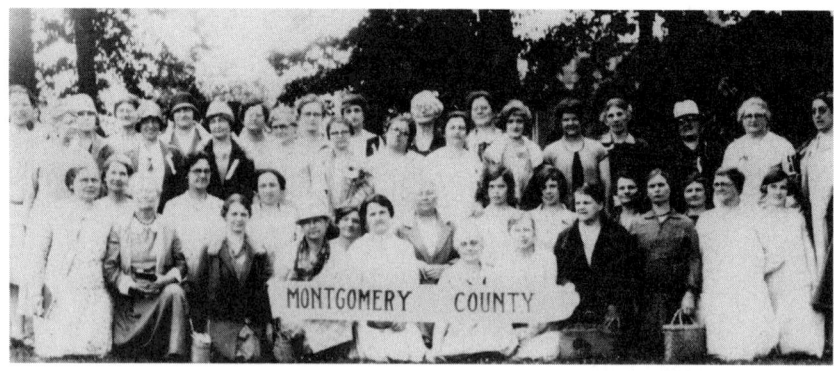

Rural Women's Retreat at the University of Maryland, circa 1928. It may be Blanche Corwin standing at far right. Courtesy of the USDA.

Chapter 1
Feed Sacks and Farm Women

Macie Schaeffer King was vibrant, energetic, lively, and enthusiastic about nearly everything. A beautiful woman, she wore her light brown hair swept up into a luxurious bun on top of her head. There was a twinkle in her green eyes and she loved to laugh. But Macie had a firmness about her that let others know when she meant business. She was a leader with a soft heart for poetry, and tried her hand at writing some herself. Macie knew how to do many things, and if she didn't, she figured it out. In 1912, when she and dairyman James D. King were married, Macie joined him in the barn and handily milked half the cows. Jim's father, Elias Dorsey King, was a prominent farmer who owned Brink, a 250-acre farm near Cedar Grove. With a happy smile, he co-signed the mortgage for his eldest son and Macie.

Macie and Jim were delighted with their comfortable, roomy house, the large bank barn,* and fertile fields near the village of Germantown. In the early years of their marriage, Jim lifted the 10-gallon milk cans onto a wagon and hitched up the horses. After a short distance, he could smell the smoke billowing from the train's steam engine. He soon unloaded at Germantown Station, glad to get his milk to market in Washington. Milk was their primary source of

---

*The bank barn was a commodious, two-story structure with a hip roof. The lower level, built into the hillside, housed the livestock, reached by a ramp formed by the hill. The upper level was filled with hay.

income, but they carted wheat and oats to Germantown, too. Next to the train station was Liberty Mill, grinding that good local grain into Silver Leaf Flour. Life was good.

By the time Macie's body began to swell with her second child, the farm was producing enough money for the farm to hire help. Houses were built for a tenant farmer and a full-time stable manager, whose wives helped Macie as part of the employment agreement. Horses were a good investment, and the Kings' beloved Percherons won prizes at county fairs. The stable manager's job included keeping the horses' hooves trimmed and shod and brushing the huge, gentle beasts until their coats gleamed.

Three children were born in the pleasant farmhouse between 1912 and 1919: James S., Helen Gertrude, and Macie Irene.

Macie's father, Allen D. Schaeffer, bought a farm on the opposite side of the road between them and lent his name to Schaeffer Road.

Both farms were highly productive and money flowed like golden cream on baked apples—until 1926. That was a bad year for Macie and Jim. Macie was ill in the hospital on the hot, dry day in August when the farm well went dry. The wheat crop was in the barn and Jim cranked up the telephone and asked the operator for Liberty Mill, intending to finalize the sale of his wheat crop. Just as he hung up, he heard a car tear down his lane and someone yelling, "Mr. Jim, your barn is on fire!"

Macie's house and twelve farm buildings went up in smoke.

Macie and Jim increased their mortgage and rebuilt.

Against all odds, their resilient spirits rebounded, too. Jim was elected as the second president of the county Farm Bureau, a powerful lobbying institution founded nationally in 1919.

Macie King purchased her first antiques and joined the Extension Service Homemakers Club with Blanche Corwin, home demonstration agent.

From a 1920s newspaper, Macie King clipped this gently nostalgic poem and pasted it in her scrapbook, permitting future readers to glimpse inside her heart.

Hang Up the Baby's Stocking
by H.M. Higgins

Hang up the baby's stocking;
Be sure you don't forget
The dear little dimpled darling!
She ne'er saw Christmas yet.
But I told her all about it
And she opened her big blue eyes
She looked so funny and wise.
Dear! What a tiny stocking!
It doesn't take much to hold
Such little pink toes as baby's
Away from the frost and cold.
But then for the baby's Christmas
It never would do at all!
Why, Santa wouldn't be looking
For anything half so small.
I know what will do for the baby,
I've thought of the very best plan;
I'll borrow a stocking of Grandma
The longest that ever I can;
And you'll hang it by mine, dear Mother,
Right here in the corner, so!
And write a letter to Santa
And fasten it on the toe.
Write, "This is the baby's stocking

> That hangs in the corner here;
> You never have seen her, Santa,
> For she only came this year;
> But she's just the blessedest baby!
> And now, before you go,
> Just cram her stocking with goodies,
> From the top clean down to the toe."

Another farm woman lived in the foothills of Sugarloaf Mountain. Margaret and Thomas Magruder once lived in the city of Washington with their two sons, born in 1915 and 1917.

Although strictly city-oriented, the Magruders liked to visit the country and were drawn to Sugarloaf Mountain. Through Margaret's sister, the Magruders met the Davis family, owners of a prosperous, historic farm nestled among Sugarloaf's foothills. Visits were exchanged and the families became friends.

As noted by Agent Van Hosen, influenza was a dangerous killer, spreading like Macie King's fire when the well went dry. In 1918, Thomas Magruder was among those felled by the epidemic. Margaret, too, was hit with the flu, but recovered and was considered immune to another attack. All hands were needed to help the sick, and Margaret quickly assumed the duties of a nurse. When she learned that the Davis family was ill and needed help, Margaret went to the country to offer aid. After bringing the family through their illness, she found that John Davis had become more than a family friend, and they were married in 1921. Bringing her two little boys, now six and four, she settled into the life of a farm woman.

Margaret Davis was a natural business manager. Farm finance is tricky, but she learned the skill quickly and assumed the responsibility of keeping farm records straight. Unschooled in dairying, she cheerfully donned the cloak of farm life and learned the intricate details of the successful farm.

Like Macie and Jim King, the Davises' major income came from their cows. Cleanliness was next to godliness, and they were exquisitely careful in achieving and maintaining the very highest standards. Their dairy barn held a wood-fired boiler that created both boiling water and steam. Bottles, buckets, and the cream separator were washed after every use and then placed in a cement container with a lid. Steam was transferred into the cement pot, sterilizing the equipment.

Many years before electricity bewitched water to rise from a deep hole in the ground, the Davises had running water. Grandfather Davis was plenty smart. He had mounted a 1,300-gallon wooden water tank on top of a small building used for nothing else and had plenty of cold water to cool the milk. Cold milk was run through the freshly washed and sterilized cream separator. At this stage in the process, cream was poured into pint bottles, milk into quarts. Molded into the glass of each bottle were the words "Mrs. John Davis, Barnesville MD Phone Buckeystown 52 F 12." Paper bottle tops were attached, each one labeled with the name of Margaret Davis.

A health inspector came to check the milk for bacteria, and every three or four months, a milk inspector came to the farm to inspect the building and equipment conditions. The Davis farm easily passed on each occasion, according to Margaret's son, Jack Davis.

Productive and successful, the Davis place was among the first farms in the area to install a bathroom. Water was pumped into the house by a windmill. Hot water came from a tank attached to the kitchen cookstove. Coils in the firebox held water, which circulated into a 4-gallon tank and was pumped to the bathroom.

Dairying provided a good life. A new baby arrived, and it was a little girl. Harriet Davis was born in 1923 and instantly became the apple of her father's eye. She was his little darling angel, and he carried her with him wherever he went. She loved riding in the wheelbarrow, so her father made a small nest for her, pushing his load to one side. Harriet laughed and clapped her baby hands, and the entire farm adored her.

One day when Harriet was eighteen months old, a vial of medicine was left in the middle of the kitchen table, seemingly out of reach of tiny hands. But Harriet climbed up on a chair, crawled across the table, and got it. Decades before the advent of childproof caps, the lid popped off and the baby swallowed the pills. And died.

John Davis never got over the death. From then on, he grieved for his lost child. Whenever he loaded the wheelbarrow, he made a little nest for Harriet.

Twins Jack and Peggy were born in 1924, and Margaret's fourth son, Leonard Isaac, in 1928.

The Health Department officer arrived at the Davis farm one day with new regulations, demanding a new dairy barn be constructed of cement blocks. Only cows could come in; no horses, pigs, chickens, sheep, or other farm livestock were allowed.

Grandfather Davis was still the owner, and he put his foot down. According to Jack Davis, his grandfather "said he was too old to build another barn and wasn't going to do it." The Davis family then could not sell milk in Washington, because they did not comply with the new regulations. No milk sales meant no money coming in. When Grandfather died, Margaret and John bought the farm from the family heirs, scraping together every cent they had.

Here's a recipe from the yellowed files of Margaret Davis:

Mince Meat
9 lbs. apples
2 lbs. currants
1 teaspoon black pepper
4 lbs. cooked meat
5 lbs. sugar
6 tablespoons salt
3 lbs. raisins
2 teaspoons cloves
1 qt. sherry wine
1 lb. suet
10 teaspoons cinnamon
1 pt. brandy
1/2 lb. citron
5 teaspoons mace or nutmeg

Chop apples. Grind meat. Add all ingredients; mix well. Put in stone jar and add more liquid when needed. Let stand a few days, stirring once in a while. Put in jars. Yield: 12 qts.

Back downcounty, close to the District Line, lived Charlotte Waugh Potter. A beautiful woman and a scientist as well, Charlotte had a master's degree in botany from the University of Minnesota. Her interest in nature enticed her into the woods, and her garden bloomed with native flowers and shrubs. The mother of two little boys, Neal and Lloyd, Charlotte was constantly busy and gave both her children a deep regard for the ethics and benefits of hard work. Her husband, Alden Potter, had a good federal job with the Department of Agriculture. But what he longed to do was to farm.

At last he was able to follow his dream. He bought a farm near the District of Columbia limits, added cows, and began delivering milk. Alden Potter, like John Davis, worried about sanitation, and Potter put his intuition to work. He invented a lid that kept impurities out of the milk bucket. Filled buckets were rushed to the springhouse and set in a trough of flowing water, cooling and keeping the milk at a constant sixty degrees.

Neal and Lloyd went along sometimes when Alden delivered milk in Cabin John. Burt and Mike were hitched to the wagon or the sled as weather dictated. Sometimes Alden also took eggs from Charlotte's chickens, her garden produce, and even her dressed chickens for market. Then came the Depression, and no one had money to buy milk. Charlotte joined Blanche Corwin's Homemakers and looked around for a way to make money.

Elsie White Daniel, a pretty woman with an oval face and dark, wavy hair, lived near Poolesville, across from an eighteenth-century farm, Inverness, built by Elsie's ancestors. Among the county's earliest settlers, the White family was well established in Montgomery County. Elsie went to a private school, the Episcopal Female Seminary in Winchester, Virginia. One holiday, she went home with a school friend from West Virginia, who introduced her to William Daniel. William and Elsie subsequently married, and Elsie was swept into a whirl of frequent household moves. William was a railroad agent, and it was his job to negotiate property for track right-of-way. Elsie yearned for Inverness. After a few years, William acquiesced and they returned to the land where Elsie was born, buying five acres and what had been the Inverness miller's house, directly across from the farm manor's gate. Elsie was content, living where her family had been for over a hundred years. Elsie joined Homemakers, and with two boys, Billy Jr. and Mansfield, and one little girl they called Boo, she had plenty to do. William bought a few chickens and soon extended his poultry flock to approximately one thousand birds. He went into business for himself, selling eggs to an established route of regular customers in the close-in town of Chevy Chase. Daniel was a charismatic fellow, and his clients liked him so well that they passed along their daughters' outgrown clothes to little Boo.

Rosa Barnes lived in an L-shaped farmhouse north of Damascus near Browningsville. The house was large. Grandparents Harriet and James Oliver Barnes lived on one side; Rosa and Herbert Barnes and their four daughters, Dorothy, Louise, Vivian, and Mazie, inhabited the longer part of the "L."

"Everybody works on a farm," said Louise Barnes Duvall, born in 1920. Grandmother, Grandfather, Rosa, Herbert, and all four children were needed. Grain was grown, but not to sell: the horses, cows, and chickens ate the grain. Like most Maryland farmers of the time, the Barnes' primary source of income was tobacco. Louise Duvall gave this firsthand account of the process:

> For tobacco the growing year begins early and February found the Barnes women planting seeds and tending seedlings. Tobacco needs new soil each year and the men worked up a plot on the edge of their woodland, plowing with the horses and raking by hand to get out all the weeds. In May the ground was warm enough to grow the seedlings to maturity. Louise, her sisters, mother and grandmother dug the tiny plants very carefully from the seedbed and Old Tet was hitched to the tobacco planter. Attached to the rear of the planter were two backward-facing seats, centered by a large water tank positioned directly above two small wheels. The women and girls took turns riding and dropping a small tobacco plant into the

holes made by the planter. As each leafy stalk was placed, the tank dropped water into the hole and the two little wheels pushed dirt around the plant. Misplacing a plant as the implement bounced along was not an option.

As the weather grew warmer and the plants matured, large, green tobacco worms threatened to destroy their crop. The children were sent to save the tobacco. Louise Duvall described the method of disposal she and Dorothy preferred.

> "When Father called, 'Dorothy! Louise! Come with me to the tobacco field,' this year we were ready," said Mrs. Duvall.
> "Father's method was simple: Just pull the worm from the plant, twist its head off, and toss it on the ground."
> Mrs. Duvall stretched her thumb and forefinger to their longest extent. "The worms were big, fat, ugly and this long," she said. "We couldn't twist off their heads. They spit at us."
> This particular day Dorothy had a pair of tweezers and Louise had a pair of small scissors. Herbert Barnes did not approve. But Rosa said, "Herbert, if those girls can't use their tools, I have plenty of work for them inside the house."
> That settled it and the girls demurely followed their father to the tobacco field, tweezers in one small pair of hands and scissors in her sister's.

Earl Linthicum hauled the hogsheads of tobacco to the Baltimore market, and when it was time for their crop to be sold, Herbert Barnes was notified and he went to Baltimore, often taking his daughters. On auction day, the auctioneer pulled a sample from each hogshead.

"If Daddy didn't like the sample, or the price, he could say so," said Louise, "and they would try again."

Potatoes were the second cash crop on their 178 acres. Dorothy, Louise, Vivian, and Mazie cut up hundreds of potatoes for planting, and every hand was needed to harvest them. Another specialized piece of farm machinery shook the potatoes as they were dug, removing the clay soil, which stuck with ferocious tenacity. Next, the girls picked up the potatoes, dumped them into a wagon, and the crop was hauled to the barn, where they were graded and tossed into 100-pound bags. It was the women's job to sew the bags shut with long needles and heavy twine.

The children went to school in the Browningsville one-room schoolhouse. Honesty is essential to Louise Duvall, and when asked about the school, she had one word to say: "Awful."

Her parents found a high school student, Lillian Gladhill, who agreed to drive the four Barnes girls to larger schools in Damascus. When Lillian graduated from high school, a younger Gladhill child drove until the Browningsville school closed and a school bus transported all of the Browningsville students to Damascus.

Early in their lives, Grandmother Barnes gave each child a Dime Saver from the Central Bank of Monrovia, telling them, "Save your dimes and then when you are old, you can live off the interest." She, Grandfather, and Herbert put all their money in the Bank of Monrovia. Their hard work paid well, and the Barnes family prospered. Happily they watched their savings accounts grow.

Eleanor Cissel Waters lived on a 525-acre farm near Germantown. Her home was built in 1815 on land deeded to William and Elizabeth Waters by King Charles in 1660 and had never been sold outside the family. Many barns and stables, five tenant houses, and several farm workers' cottages gave the farm the appearance of a small village. According to family legend, the main house was built from bricks carried as ballast in an English ship. Eleanor, a former schoolteacher, was a traditional farm woman, happily engaged in a prosperous dairy business.

Ida Mae Henderson lived on a 400-plus acre farm near Travilah, raising corn, wheat, hay, and potatoes. She and her husband were sharecroppers, meaning they neither owned nor rented, but did all the work and depended on the landowner to share his largess. The use of their house was part of their wages. However, the life of sharecroppers was precarious. Should the owner decide he wanted them out of the house, he could evict them with no notice. If the man of the household died, the family could be and usually was evicted immediately; after all, the landowner needed the house to attract workers.

Eventually the Hendersons were able to rent a farm and moved to land near Potomac. There were twelve children, six girls and six boys.

Ida Mae watched her husband become disabled. Her sons took over their father's work load, attending school when they could. Ida Mae worried about paying the rent and caring for her children. She could make their clothes, but where would she get the money for shoes? Soon, Ida Mae joined Blanche Corwin's Homemakers Club.

Part of Blanche Corwin's job was to call on the farm women of Montgomery County, spread out between the Potomac and Patuxent Rivers, Sugarloaf Mountain and the District of Columbia. As home demonstration agent since 1922, Blanche made it her business to know these country dwellers and to help them. She loved her work and knew she was good at it. A professional job such as hers was hard to get, and only a woman of rugged, strong personality and faultless character would be considered for the position.

Regulations were strict. A home demonstration agent was not allowed to marry. However, her salary gave her economic independence, making marriage an option rather than a necessity. She could not wear gold bracelets or gold rings. Her business was to educate farm women, and she was required to know all the home arts, including tatting, knitting, crocheting, quilting, embroidery, spinning, upholstery, chair caning, mattress making, rug braiding and hooking, and safe cooking and preservation of food, as well as maternal and child care. Blanche Corwin did know all these things.

Blanche's job title was assistant extension agent, chosen by Wardney C. Snarr, who answered to Dr. T. B. Symons of the University of Maryland. Symons's position in the land grant college made him administrative chief of extension work in Maryland.

Dr. T. B. Symons saw to it that each extension agent obeyed the rules. The home demonstration agent had to have a college degree. She was required not only to acquaint herself with the farm women of her assigned county, but also to organize Homemakers Clubs. In addition, she had to assume a leadership role in the Rural Women's Retreat held at the university campus each summer for Homemakers Club members throughout the state.

Corwin owned a car, but it's likely that when she was hired in 1922, she used a horse and buggy to get around the rural areas and probably stayed overnight at the farm of a Homemakers Club hostess.

Serious, but with a twinkle in her eyes behind rimmed glasses, Blanche was deeply committed to her work and was knowledgeable on a wide variety of topics. She dressed simply and was liked and admired by the farm women, teaching them all sorts of skills.

The year Blanche became assistant extension agent, 1922, was an exciting year for women, just two years after the nineteenth amendment to the Constitution was ratified and women finally could vote.

Chapter 2
Upheaval

During the 1920s, farming provided a good life for farm women. By the end of the decade, dairy farming replaced tobacco as the dependable golden goose of most farms across the country. Then came nationwide drought and, following close upon its heels, the collapse of the stock market. From 1929 to 1932, net income for farms in the United States fell a whopping 70 percent.

Farmers reacted with desperation. They threatened to hang a federal judge, overturned milk trucks, picketed packing plants, and boycotted farm sales.

Some distributors offered to buy farm products, but at a price below production costs. Farmers poured their milk on the streets of Washington in protest. No longer friendly and amiable, dairymen turned ugly toward one another. On the local scene, bottles marked "Mrs. John Davis" were smashed by competitors.

Montgomery County farmers devised a plan. They organized. The Farmer's Non-Partisan Taxpayers League tried to convince legislators to lower taxes and reduce county spending during this emergency. Any further bond issue was opposed, pointing to the county debt. When the county commissioners proposed a $300,000 bond issue, an overwhelming majority of the Sandy Spring Farmers' Convention voted against new bonds for any purpose. The county backed off, but bad times did not go away.

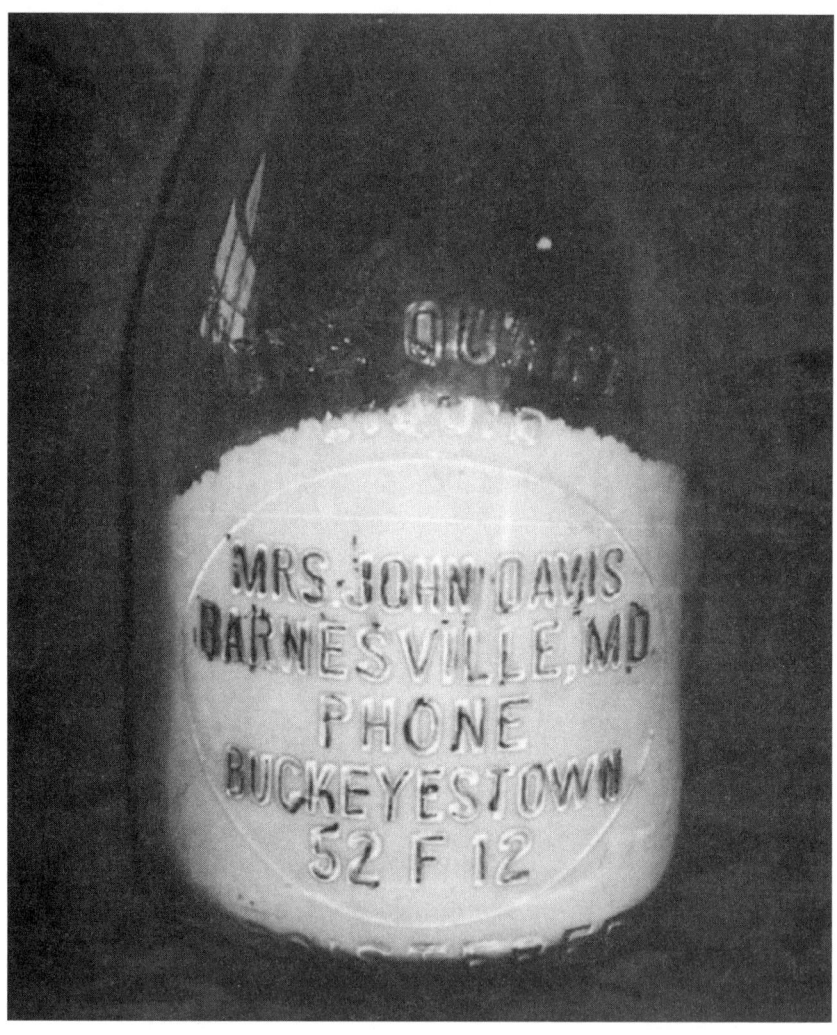

One of Margaret Davis's milk bottles escaped savagery and is preserved and treasured. Circa 1935. Courtesy of Nona Schwartzbeck.

The Social Service League, led by Dr. Jacob Bird, ran out of money. As the Depression deepened, Dr. Bird said that 50 percent of Montgomery County's destitute constituents would have to be turned away due to lack of funds. Hollow-eyed families lined the courthouse steps. The local newspaper printed a plea for clothing. More than 150 children were in need of "every conceivable garment," new or used. Readers were urged to leave donations at any police station. Canned food was needed, too, and canning jars were available for free.

Farm women had no money to buy new clothes. Adults wore their old dresses, but small boys and little girls found that last year's pants and pinafores didn't button anymore. In response to needs of the Depression, feed and flour mills began packaging their products in bleached cotton sacks printed with small, colorful designs. Red tulips were popular; little anchors, leaves, stripes and other patterns marched across the fabric. Women accepted the sacks with delight, and with their usual ingenuity put them to good use. When more feed or flour was required, they chose the bag with the prettiest print. Sometimes the feed sacks were made into beautiful quilts. In work-worn hands and flying fingers, feed sacks became aprons, dresses, and bonnets for women and girls and shirts for the men and boys.

"They were just the prettiest things," said Peggy Kingsbury, remembering the feed-sack clothes made for her children by her neighbor, Sarah Wade.

Peggy's children were lucky. The unlucky children were those whose parents could not feed them and abandoned them at an orphanage. Sometimes these children were put on trains and sent into the country in hopes that they would find people who would take them in and feed them. At towns along the railroad, these Orphan Trains were met by people who agreed to take a child. More than a generation after slavery was outlawed, little children were paraded off the train to be examined by a group of prospective foster parents. They might or might not be chosen, and those who were not climbed back on the train and repeated the experience. Those who were chosen were sometimes the victims of sad consequences.

For the helpless children of the poor, the Depression was inexpressibly painful. This experience from the 1930s was told to me by a friend who asked to remain anonymous:

> When I was five years old, my mother died. My sister was three and my brothers were seven and nine. My father took us to downtown Rockville and told us to wait at the corner while he went to Murphy's to buy ice cream.
>
> We never saw him again. All day long we stood at the street corner. A policeman stopped and asked what we were doing and we told him we were waiting for Daddy to bring us ice cream. The policeman kept an eye on us. Toward the end of the day, the policeman took my hand and my little sister's hand and walked us to the orphanage. Well, maybe it wasn't an orphanage, but a county building of some sort. They asked us to name our relatives. We didn't know their last names, but we knew we had an uncle who was a fireman and we knew his first name. They found a [man with that name] at the Kensington Fire Station and it was him all right. But he didn't want us. My aunt wanted to take us in, but he didn't. So they made some phone calls and the policeman took our hands again and walked us to a house where a woman opened the door.
>
> The policeman said, Meet your new mother.
>
> This woman was so mean to us. When she moved to a farm near Clarksburg, we did the farm

work. We learned to clean, cook, milk cows, shovel manure and lift ten-gallon cans of milk from the springhouse to the milk truck. When I was eleven years old, that was my job.

If I didn't move fast enough, the woman would hit me. One day she hit me in the head with an iron skillet, so hard I am deaf to this day. Health care? The county had a dentist who came around to the school. He made me lie down on the floor, put his knee on my chest, and pulled my teeth.

Blanche Corwin, knowing all this, seeing all this from her office in the Rockville courthouse, felt compelled to do whatever she could to help. She saw the families at the welfare office and could scarcely have missed seeing the abandoned children, too. As she connected with extension agents in other areas, she heard about farmers' curb markets being set up in the southern states.

Corwin conceived a plan. She would try something that had never before been done. She began to teach members of her Homemakers Clubs about marketing, setting prices, maintaining high standards, and being the best at what they did. These were perilous times indeed, and Blanche Corwin knew they called for desperate measures.

Outside the city, farm women worried. They could grow their own food, stitch clothing from feed sacks, and can fruits and vegetables for the winter. Macie King could quit buying antiques. Eleanor Waters could give up the hired help. Charlotte Potter grew all her family's food. But what about the mortgage payments? And taxes? And shoes for the children? And college for the older ones? And Grandfather's medicine? Who would feed their old retainers if their servants were turned out of the only place they called home? The women all knew of farms that were foreclosed and auctioned off on the courthouse steps. In this depressed market, there was plenty of opportunity for those few who had cash to buy, and the bids were very low.

Life wasn't so good anymore.

The Central Bank of Monrovia locked its doors, never to reopen. All the Barnes' family money, the girls' Dime Savers—everything was swept away like West Texas dust. There was a market for the Barnes' crops, but no buyers had money to pay, and the cash received was less than the actual cost of the seeds.

"It was the only time I ever saw my father cry," said Mrs. Duvall, shaking her head.

Rosa Barnes heard of a way out of their desperate situation. She joined the Damascus Homemakers Club, met Blanche Corwin, and began to learn marketable skills.

"Aunt Blanche could do anything," said Mrs. Duvall. "If she didn't, she found someone who did, and this person taught the others."

Appreciation gave birth to affection, and, to the children of her Homemakers Club members, Miss Corwin was indeed "Aunt Blanche."

Chapter 3
The Montgomery County Farm Women's Cooperative Market, Inc.

Blanche Corwin knew of the precarious situation in the rural community. She knew that farms were mortgaged and taxes were due. She knew that farm income had been cut off in some cases and in others, lessened considerably. The women in her Homemakers Clubs were growing desperate. Nearly all of them lived in homes that were mortgaged, and they and their children could become destitute. Bad times called for heroic measures, and these women, Corwin was certain, were capable of doing almost anything.

Blanche knew that Montgomery County was in an advantageous position. Since the Civil Service Act of 1883, federal workers in the capital city could keep their jobs through changing administrations. Fairly immune to the Depression, their regular paychecks meant that many Washingtonians would not be directly affected by the economy. Coupled with these matters, Blanche knew that each club member could bake, cook, and sew to perfection.

Why not put these two situations together to benefit both city people with paychecks and farm women without?

A market! It would have to be not only good, but the best, and it would need a special twist. A market owned, managed, and staffed exclusively by women! The market would have to be in the best location to attract city people

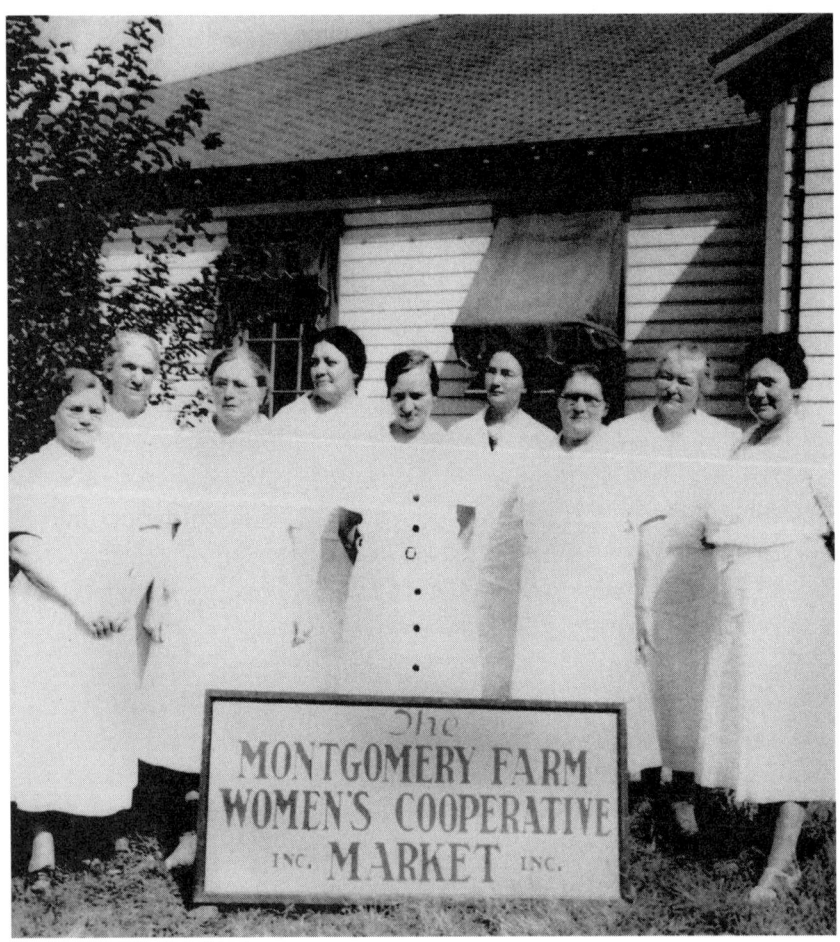

The Montgomery County Farm Women's Cooperative Market Inc. Right to left are: Macie King, Mrs. Tarent, unidentified, Margaret Davis, Mrs. Stanley, Eleanor Cissel Waters, Mrs. Forest King, Rosa Jones, Julia Williams. Courtesy of King Barn Dairy MOOseum, identifications by Jack and Elizabeth Davis.

and (especially) their money. Bethesda, bordering the District of Columbia, would be perfect. Although curb markets were somewhat successful elsewhere, no one had tried a market run solely by women.

Could the farm women undertake yet another Herculean task? Even in good times, women worked longer hours than the men, yet were financially dependent on their husbands. But Blanche knew these women and their capabilities. She yearned to help as she worried and wondered about her embryonic idea. Could the farm women get to Bethesda? The distance between Sugarloaf Mountain and Bethesda was at least fifteen miles as the crow flies, and she knew the crow didn't fly the winding, mud roads in the rural areas.

Macie King and her sister-in-law, Pearl King, drove their own Pontiacs, but Margaret Davis and a few others did not drive and were brought to club meetings by their husbands or sons.

Tentatively, Blanche Corwin put forth her original idea at a Homemakers Club meeting. The farm women would open their own market. Utilizing the skills they had learned, they would bake, cook, butcher, braid rugs, quilt, and so forth, and exchange their hard work for cool cash. They would sell at a rate slightly higher than regular stores, but they would produce a premium product. They would establish a market site and organize a cooperative that would be owned and run entirely by women. With no middleman, profits would be high.

At a time when farm marketing in Maryland was done by men, and women were expected to stay at home and keep house, this was a shocking idea. But their economic situation was growing more desperate every day. And the women knew and trusted Blanche Corwin.

The club members agreed that with streetcars racing back and forth from Washington, creating easy access for those government paychecks, Bethesda was ideal. Furthermore, foreign embassies and the wealth of the world were within a short drive. The women would rent space at first and find something permanent when they became established. The perfect site, Corwin proposed, was near the tree-lined neighborhood of Edgemoor, and near Edgemoor was a point of land that would be just right for a permanent market building.

Restless hands lay still as the ladies leaned forward to hear better. Suddenly they burst into excited chatter. A market of their own! What would their husbands say? Was it possible that a group composed only of women could buy property on the edge of an exclusive community, own a store, and manage it? Without the help of men?

Of course it was! Blanche Corwin assured them.

As time went on, the women discussed the products they would offer. Blanche Corwin stressed that only the very best, freshest foods would be accepted, and they must adhere to strict regulations. For two years, Blanche Corwin focused on teaching the women about marketing, setting prices, maintaining high standards of cleanliness, and being friendly and respectful to their customers.

Abandoned stores were easy to find during the Depression and, while she worked on the best site for a permanent structure, Corwin rented a small building on Wisconsin Avenue, unfashionable but available. The women put out flyers and notices. They told their friends, their church groups, the press.

On February 2, 1932—a bitterly cold morning—nineteen farm women arrived in Bethesda very early. Shivering in below-freezing temperatures, the women set up card tables and unloaded their cakes, bread, poultry, sausage, jams, and jellies. Husbands came, helped set up, and left. For some of the men, accustomed to supporting their families and dismayed at their inability to fulfill their role, it was hard to relinquish their position as family breadwinners. A few supported the fresh ideas, but others mocked their wives and predicted total failure.

As the building warmed up, the ladies took off their coats to reveal sparkling clean white dresses. They pinned on white lace caps. Precisely at 7:00 a.m., Blanche Corwin yelled, "Okay! Quiet now!" She held up a small bell, which she rang as she opened the door.

A flood of people poured into the store. Kept too busy to notice, the nineteen farm women and one smart home demonstration agent sold almost everything they had. When Blanche Corwin rang the closing bell and shut the door, they looked around them and saw empty tables everywhere.

There was joy at the farms that night.

By June, thirty-one farm women were taking their wares to Bethesda. Corwin secured an empty lot on the corner of Wisconsin and Leland streets and set up a large tent, 45 by 90 feet. The women made their first group purchase, an ice box and 600 pounds of ice. Nearly everything they could bake, grow, or slaughter sold.

Back taxes were paid. Mortgage payments were caught up. Children who had been kept out of school because they had no shoes trudged off to learn their ABCs. Their older siblings looked at college catalogs.

"We got a bathroom," said Elizabeth Daniel Davis.

Margaret Davis began to pay off the mortgage, and John was delighted, according to their son, Jack. She set a jar on the kitchen table and dropped in all the 50-cent pieces she earned. "For a car," she announced to her startled family.

Ida Mae Henderson, small but strong-willed and determined, organized her six boys to do the outside work and her six girls to do the chores inside, providing Ida Mae with time to bake and cook for market.

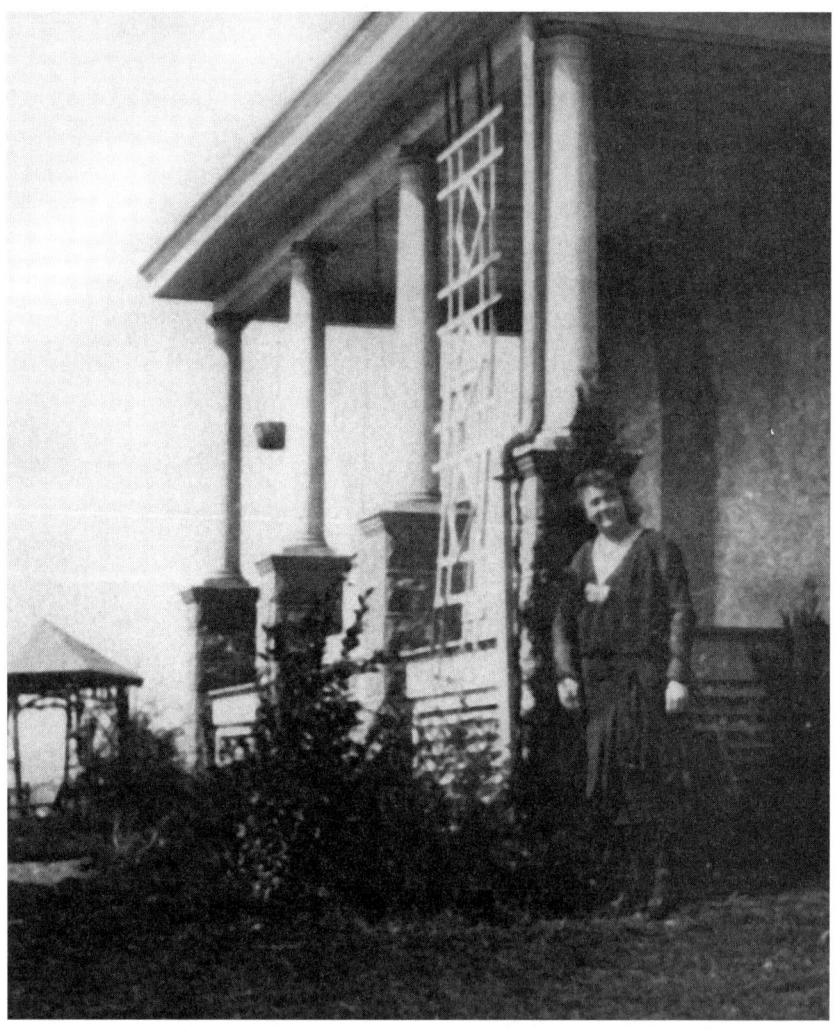

Macie King at her home, circa 1936. Courtesy of Garner W. Duvall.

Macie King got two new, modern stoves and set one in the kitchen beside the old woodstove and one in the basement, where a second woodstove kept the rehired farmhands warm as they ate. She appreciated Jim's strong support and reliably constant help with her marketing. She watched as he unrolled blueprints for a new dairy barn.

Rosa Barnes took each of her girls in turn to help her on market day. She drove the family auto to Bethesda and on the way picked up a neighbor who was in a similar financial situation.

"Our weeks became centered around the market," said Louise Duvall, happy laughter sweeping into her voice.

"Mondays we washed. Tuesdays we ironed. Wednesday was our day to do all sorts of things—clean, shop, get haircuts, or whatever we had to do. Thursdays we baked cakes, usually Mother's special coconut cream with seven-minute icing. Fridays we dressed chickens. And Saturdays we went to market."

The Great Depression was a leveler; women were seen in a different light. They could and they did own property and provide income. Though there were some exceptions, many of the husbands were delighted to turn over their role as breadwinners to their wives.

However, local politicians worried. Farm women were assuming leadership roles and organizing. What was next, a union? Socialism? They gave women the vote and look at what they want now. Complete breakdown of the traditional family? The Great Depression had spread throughout the world, and established governments were falling. Politicians grew wary. Old ways were being cast aside. And in Germany, fascism was taking over and little boys were doing the goose step.

America worried, listening to their crystal sets for more bad news. War was on the horizon, and women were getting entirely too uppity.

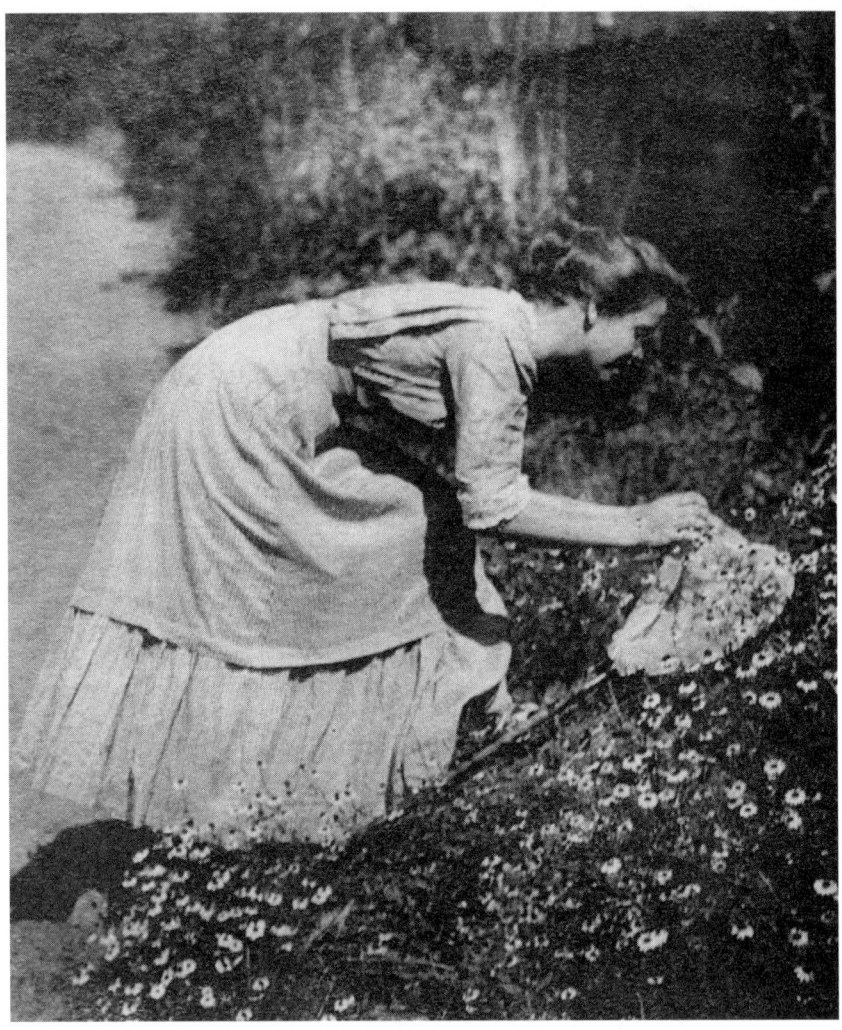

Charlotte Waugh Potter identifying insects on her wildflowers. Courtesy of Neal and Marian Potter.

In 1930, unionizer Mother Jones, celebrating her 100th birthday in Silver Spring, had said, "A wonderful power is in the hands of women."

Her words hovered above Bethesda, settling on all the people as surely as the scent of warm, spicy cinnamon rolls invigorated that first cold February morning at the market. The word "women" coupled with "power" made women shiver with joyous excitement. Some of the men, however, quaked with fear.

The Montgomery County Farm Women's Cooperative Market was such a success that the women were rapidly outgrowing both the store and the tent. Blanche Corwin worked on articles of incorporation and looked for a larger site. The more she looked, the more brightly gleamed Edgemoor Road. There was an empty Piggly Wiggly Store where the market ladies could set up right away.

Absolutely no way, said the property owners of Edgemoor Road. They were joined by the powerful head of the dominant Democratic party, E. Brooke Lee, ear bent to his constituents in Edgemoor. As Edgemoor protested, the county politicians, already on pins and needles, listened. The editor of the Rockville newspaper hinted that Edgemoor was sacrosanct and Blanche Corwin was not to come near. The farm women could "not secure the Edgemoor site for a farm market and repeated agitation on the part of their leaders may bring about official action from the county authorities," wrote a journalist.

The whole idea was preposterous. Farm women "at the entrance to one of the most beautiful and desirable residential sections of the metropolitan area"? Miss Blanche Corwin, continued the local newspaper, *The Sentinel*, has been reprimanded and "told that continued agitation on her part may bring about withholding of her pay."

Blanche Corwin, seeing that the market ladies were pulling out of their despair, committed herself not to quitting, but to further strengthening the market. Tension mounted. Farm women were in the limelight, but this was not the attention they wanted. As the crisis peaked, Dr. T. B. Symons sent Blanche Corwin a telegram. After that, he called for a meeting to be held at the courthouse on a Monday night. County politicians came. Extension Agent O. W. Anderson came. Edgemoor residents showed up en masse, but not many farm women could get themselves to Rockville on a Monday night.

Blanche Corwin was accused of spending too much time on the market at the expense of her other duties. She had been fired in the telegram, Symons announced. He then introduced Miss Edythe Turner as the new home demonstration agent.

Blanche Corwin walked out, followed by a few of the market women who were able to attend the evening meeting. The larger group stayed with Dr. T. B. Symons, O. W. Anderson, E. Brooke Lee, and Miss Edythe Turner.

"Farm Market in Definite Break After Stormy Meet," screamed *The Sentinel*.

Using her own funds, Corwin rented the Edgemoor Piggly Wiggly building and also a market in Takoma Park. Charlotte Potter and a few of the other women followed her, but without the support of the extension service, the county politicians, and the press, Corwin's group found their markets could not survive. Potter rejoined the cooperative.

Blanche Corwin's idea flourished. But Blanche Corwin herself disappeared from historical archives.

Chapter 4
Fame and Fortune: The Only Cooperative Farm Women's Market in the World

"We have saved our farms!"

Although politically bumpy, the Montgomery County Farm Women's Cooperative Market was an instant success.

Macie King was elected as the organization's first president, and she accepted the challenge with her usual graciousness. Along with the professional assistance of Miss Edythe Turner, Macie took the reins of leadership and the embryonic business sprang ahead.

All summer long, fragrant odors of fresh sausage and herbs, homemade breads, lavender, and rosemary blessed busy downtown Bethesda and enticed buyers inside the tent. Fresh-cut gladioli splashed lavish color, lending a gardenlike ambience. Business boomed, and there was no room for more tables. By the end of November 1932, there were forty-two ladies in white and thirty on the waiting list.

Some of the women found the work too strenuous and dropped out. Some would-be vendors were rejected. The agreement was that no one would be accepted without a visit to her farm, and a few kitchens were determined to be unsanitary. Some of the women had no floor, but hard-packed dirt could be swept clean and was not an automatic out. Sure to be expelled were women who bought products elsewhere and tried reselling them at the market. But those

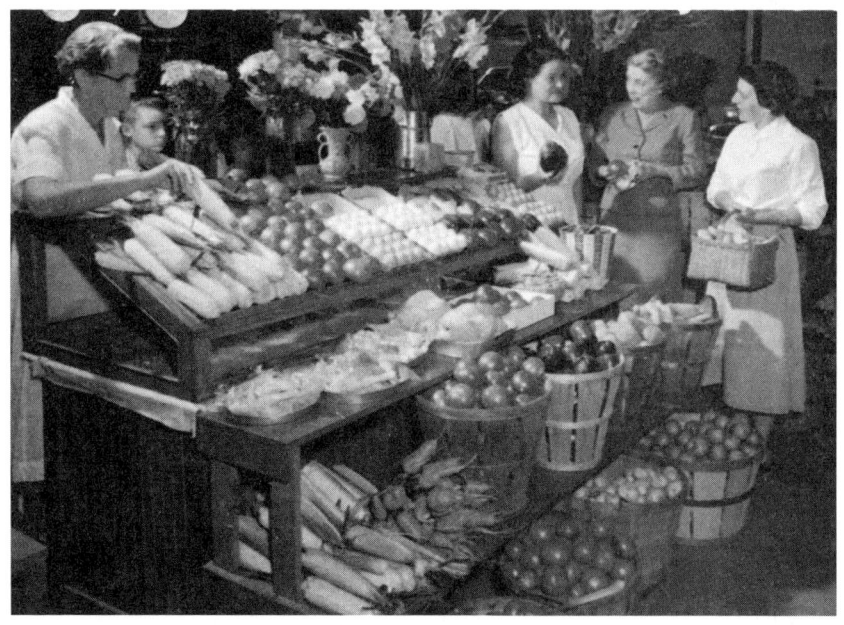

Farm Women's Market, 1957. Courtesy of Brook Photographers and of Elizabeth Davis, who saved the postcard all these years.

who were willing enough, strong enough, and determined enough settled into their role, worked very hard, and brought home real money. They produced the family income and stood tall and proud.

In the meantime Leon Arnold, the owner of the lot on Wisconsin and Leland, was not unaware of the excitement going on right under his nose. Behind the tent, Arnold constructed a long, low building, esthetically pleasing to the eye. A gable-roofed center bay protruded a few feet from two flat-roofed, matching wings. Wide, double doors split the central bay in a graceful arch. No one could miss the entrance.

Arnold met with Macie King and Edythe Turner and offered to rent them the building at $125 a month.

It was time for a special celebration. In the grassy yard near the board fence, the women erected a wooden sign that read, Montgomery County Farm Women's Cooperative Market. On Friday evening, December 2, 1932, Edythe Turner opened the double doors of the brand-new market house, and a crowd of one thousand people came in.

The building seemed to exude a welcome. Windows along the front facade gazed out at Wisconsin Avenue with a cheerful twinkle provided by newly installed electricity. Some of the windows sported an awning, and those that did not had crisply ironed organdy priscilla curtains, tied at the middle with a perky sash like a little girl's pinafore. A wide cement walk led visitors from the main street to the entrance. Everything about the structure sang of open green spaces, gardens, charming farm homes, a warm, smiling welcome, and plentiful food.

Macie reminded the ladies of proper first appearances. White dresses and little white hats were required. In winter, when the customers came wrapped in fur-trimmed wool coats and felt hats, a colored wool sweater was acceptable but not encouraged.

Prominently hung where both buyers and sellers could see it was a list of the day's prices. No one undercut her neighbor; price competition was forbidden and "considered unethical and dangerous to the success of the project."

Market stalls were about five feet long and could be divided in the middle for two vendors. Once they had electricity, the farm women bought electric refrigerators for their stalls. Before each market day the refrigerators were washed with delightfully hot, running water. Business was excellent, and none of the women wanted ever to move their market again.

At the end of 1932, all the state requirements were fulfilled and the Montgomery County Farm Women's Cooperative Market Inc. was granted a charter under Maryland law. The women had subscribed to capital stock, written bylaws, and approved articles of incorporation. The market got a telephone, OLiver 2-2291. Macie passed the reins of leadership to Mrs. Chester Clagett in January 1933. Clagett would handle the post for two years.

The articles of incorporation laid down strict rules for membership. (1) Women only. (2) Market women must belong to an Extension Service Homemakers Club. (3) All products sold would come from the seller's own farm of six acres or more; she lived on this farm and the farm was her primary source of income. (4) The highest standards would be achieved and maintained. (5) Each member was to purchase two stock certificates at $2 a share and pay 5% of each day's sales as well as $2 a month rent to the association. This sum would cover costs.

Life changed dramatically for the market ladies. Their reclusive time was over. No longer did they work on the farm all day, every day, with church on Sunday. Now they worked twice as hard on the farm, kept an even larger garden, baked, cooked, dressed chickens, ground mincemeat and sausage, and spent long hours in Bethesda on Wednesdays and Saturdays. This labor was added to their regular job of keeping their homes clean and their husbands and children fed—and church on Sunday.

For some families, tradition dictated getting together weekly. Margaret Davis expected an extra ten to twenty relatives to drop in every Sunday for dinner. Still she managed to bake forty to fifty loaves of bread and twenty to twenty-five dozen rolls for market twice a week as well as fill the farm truck with garden produce, hams, and sausage. She no longer thought of herself as a city girl, but was too busy to notice whether she was having fun or not.

Another market lady, Pearl King, made ninety cakes on Tuesdays and another ninety on Fridays.

Elsie Daniel took to market her coconut, angel food and devil's food cakes; her dressed chickens; eggs; mint, zinnias, chrysanthemums, coreopsis, and larkspur; and her specialty, jellied chicken in a mold.

Charlotte Potter took dressed chickens, lemon tarts (her specialty), filled cookies, Valley Forge Bullets (another specialty cookie), molasses cookies, plum pudding and, in August, fresh corn. When the corn was ready, Charlotte and her two sons were up at 4:00 a.m., "wet pollen all over us," said Neal Potter. Their corn was reputed to be the absolute best, and "people lined up to buy." Charlotte took orders for firewood, and in the spring she brought flowers. Her handwritten account lists tulips, chionodoxa, lily, iris, amaryllis bulbs, erythronium, crocus, scilla, bleeding heart, Jacob's ladder, primroses, lilac, and sweet woodruff. The botanist in Mrs. Potter loved flowers.

Macie King filled her basement kitchen with goods for the market: cakes of all kinds, jellies and jams, sauerkraut, and sausage she made herself. Somehow Macie found time to enlarge her field of interest. She joined Questers, a club for women interested in improving their field of knowledge, gave up canasta for bridge and then taught bridge, lectured on antique glass, corresponded with antiquers across the nation, hooked and braided rugs, and joined Jim on his trips to Chicago for Farm Bureau business.

Fame was spreading. General Eisenhower placed an order for Macie's hominy grits, and the transaction attracted newspaper attention. Macie looked farther afield for new products. Driving into the Catoctin Mountains, she bought wooden crates of berries and made jellies and jams. She was awarded a license as a government-approved slaughterer. Macie's baked beans and sauerkraut were a big hit, and she began making them in 5-gallon crocks. She had the back seat of her Pontiac removed to make room for her specialty cakes and crocks of beans and 'kraut.

Then she accepted a boarder.

Twelve-year-old Frances Wilmot thought she would die if she didn't have a horse. Her father worked for the federal government and was not about to move out in the country beyond Bethesda. Meanwhile Frances pined and dreamed. Finally her mother had an idea. Living near the farm women's market as she did, Mrs. Wilmot walked inside and looked around. The first woman she encountered, at stall number one, was Macie King. She struck up a conversation and liked Macie at once. So she confided her problem. Did Mrs. King know of anyone who had a nice, steady horse that she could rent for her young daughter to ride? Macie thought for less than a minute and invited Frances to come home with her. Frances could spend an entire day in the country, riding Macie's own mare, Lady. Meals were included and sometimes overnights, too. All for $1 a day!

By tradition, everyone on a farm had a job, and Macie knew just the thing for Frances: rounding up the cows for their evening milking. (Sixty years later the memory

𝔓ursuant to the authority vested in me by the PRESIDENT of the UNITED STATES acting under the SECOND WAR POWERS ACT, 1942, passed by the CONGRESS of the UNITED STATES on March 27, 1942.

Mrs. James D. King

is hereby 𝕷𝖎𝖈𝖊𝖓𝖘𝖊𝖉 𝖆𝖘 𝖆:

𝕮𝖑𝖆𝖘𝖘 2 𝕾𝖑𝖆𝖚𝖌𝖍𝖙𝖊𝖗𝖊𝖗

and is authorized to conduct operations in accordance with all orders of the War Food Administration issued for the purpose of distributing meat in the interest of effective prosecution of the war. Issued this the 13th day of September, 1943

ACCEPTED BY:

NAME

TITLE

WAR FOOD ADMINISTRATOR

REGIONAL DIRECTOR
FOOD DISTRIBUTION ADMINISTRATION

License No. 51-015-P-7

Macie King becomes a licensed slaughterer, 1942. Courtesy of G. Duvall.

made Frances Wilmot Kellerman laugh. "Those cows knew what to do," she said. "They didn't need me, but it was such fun doing it.")

To many of the vendors, the cooperative was enlightening. On market days, they saw a way of life far different from rural Montgomery County. Staff from foreign embassies, wealthy Washingtonians, and even the first lady of the nation leaned across the showcase to purchase the last dozen rolls. On Wednesdays and on Saturdays new fashions and the latest styles paraded before the women's weary eyes. In the shops along Wisconsin Avenue, they saw refrigerators and electric stoves, elegant clothes, and shiny new cars.

The Davis family had never experienced hunger, but meals became more varied. When Mother went to market, said Jack Davis, "We had steak for breakfast. And we all started dressing better."

Elizabeth Daniel got her first brand-new, never-worn coat. It was a beautiful blue-and-red plaid that she and her mother saw in the window of Woodward and Lothrop, next door to the market. They went inside the glamorous, glittering building; Elsie handed over her market money and placed the bright coat across her daughter's shoulders.

Then Leon Arnold received a good offer for his attractive new building and decided to sell it. By this time, September 1935, Eleanor Waters was president of the association. She conceived a plan: the market women would buy the building.

Some of the ladies were aghast. Women buying commercial property?

Everyone knew that land records, deeds, and mortgages were always in the name of the head of household, and this person was always the husband. Women were referred to solely by their husbands' names, even by close friends. They could not serve on a jury, and still had limited rights.

Eleanor—Mrs. Julian Boyd Waters—was not a slave to established custom. A tall, strong, attractive woman with a determined set to her mouth, she, Macie King, and perhaps others drove to the Cooperative Bank in Baltimore and asked Frank Bomberger, the bank president, for a loan of $50,000—the equivalent of $750,000 today.

What collateral do you have? Mr. Bomberger demanded.

About thirty-five really energetic farm women answered Mrs. Waters with a smile. The banker intimated that this was not the kind of collateral usually accepted in his business.

"The bank officials thought I was crazy," Eleanor said later.

Frank Bomberger was adamant; farm women could never earn $50,000, nor could a business succeed with women only. Sixteen years earlier, women couldn't even vote.

Undeterred, Mrs. Waters invited Mr. Bomberger to visit the market.

Two weeks later, as she was busily selling her regular supply of milk, butter, cream, cheese, eggs, sausage, headcheese, hams, spare ribs, chickens, jams, and jellies, Eleanor looked up to see the banker standing to one side, observing the crowd of eager buyers. Their eyes met and he came to her stall.

"You may have the money," he told her.

And the handsome new market building and its ideal market location was theirs.

Along with the carrot came the stick, and the women found themselves facing a $50,000 debt. More cakes and pies and homemade sausage and piccalilli relish and sauerkraut and baked beans and potato salad! More gladioli and iris and zinnias and chrysanthemums. More cabbages! More Valley Forge Bullets! The crowds continued to grow. Embassy limousines lined up outside. As farm women's food was set on the best tables in Washington, the White House got a whiff of all that good cooking. Rumors spread north on

Wisconsin Avenue faster than the big, black Packard could roll, and Eleanor Roosevelt herself came shopping.

Eleanor Waters told a reporter about an unidentified market participant. This woman and her husband bought a farm at the peak of the 1920s financial boom, Waters recalled. They purchased chickens and blooded cattle, all on borrowed money. When the crash came, they sold all their livestock at sacrifice prices. Then the woman joined the Farm Women's Market.

"That woman has re-purchased her herds, paid her bills and is supplying some of the best butter, eggs and cheese . . . in the stalls," said Waters.

In 1937, Woman's Home Companion sent staff journalist Anna Richardson and a photographer to document the wonder. The Montgomery County Farm Women's Market, according to the Companion, is "as distinctively American as the Lincoln Memorial and the Washington Monument . . . . Thousands of Washingtonians visit it every week and hold it in warm regard."

She described the view inside as four rows of counters and showcases facing the visitor, and "behind each counter stands an American farm woman dressed in a white uniform. Instantly your imagination leaps to lush rolling acres, well-fed stock, contented poultry, carefully tended truck fields and, best of all, a sunlit kitchen.

"In spring and summer the stalls are banked with daffodils, iris, apple blossoms, peonies, roses, and stock. In the fall chrysanthemums and autumn leaves salute you with spice odors, while nests of moss, tiny fern, and checkerberry invite you to fill a glass bowl with woodland trophies that will thrive the winter through." There were hominy grits, gingerbread, piccalilli and corn relish, smoked sausage, spiced meats, just-baked bread and fresh doughnuts, and "chicken as golden yellow as a spectator's blouse at a tennis match."

Richardson described golden butter, perfectly matched white eggs, pound cake and yellow cake, six varieties of sausage, "and in the center glistens a cake of ice in a glass bowl." But Richardson didn't forget it was the Depression, after all, when one hand must always be on your purse. "Oh yes, you have to pay for perfection, but ask any Washingtonian whether it is worth the extra charge."

An unidentified vendor told a reporter, "I may dress as high as fifteen frying chickens for one market day. And for the Christmas Eve market I dressed turkeys from 7:00 a.m. to midnight. At 2:30 a.m. we packed them into our truck."

Eleanor Waters increased her livestock to 2,000 chickens, 110 hogs, and 400 turkeys. One hog was slaughtered and dressed each week for the Wednesday market, and two for Saturdays. Not content with the livestock alone, Mrs. Waters grew corn, tomatoes, cucumbers, and cabbages, and also baked for the market.

Farm women were more tired than ever.

Before electricity came to the country, a few of the farms had an independent source of running water and electrical power. Louise Barnes Duvall described the windmill and Delco Battery Plant her father set up at their farm in Browningsville, Maryland.

Before 1920, the year Louise was born, Herbert Barnes dug a well on his farm's highest hill, and installed a windmill on top of it with pipes running to the house. By gravity flow, a tank in the house filled with water. Hot water came from a tank beside the wood range, 5 to 6 feet tall and 2 feet wide.

"On cold days, I loved to back up to that water tank," said Duvall. "Ummm, it was always toasty warm."

Upstairs was a bathroom that had plenty of hot water, although how the water got upstairs remains a mystery.

For electricity, her father built a small shed, "about the size of a privy," to house a Delco battery plant. The shed measured about 4 by 5 feet, and was built onto the end of the meat house. Inside, one wall was banked with two or three rows of glass batteries, each battery measuring about 1 foot wide and 1.5 feet tall. Inside the batteries was a fluid of unknown composition. In one of the batteries was a glass tube about 8 to 10 inches long, with a ball on the end of the tube. The ball indicated the amount of "power," Duvall said.

Beside the batteries was an engine on a cement block. It was Louise's job to keep the engine filled with kerosene, the level of which was indicated by the ball. If the ball was not too low, a button could be pushed to restart the engine; if the kerosene were depleted below a certain level, the engine was started with a crank like the one used on early autos. Wires ran from the shed to the house. Each of the thirteen rooms had lights. The voltage was wrong for large appliances, but the family did have both a clothes iron and a fan operating off the Delco plant, and a pump to carry water to the bathroom upstairs.

Self-sufficient and independent, the Barneses had a good life. "Grandmother said to always build a house on the worst piece of ground and save the best land for growing crops," said Louise Duvall.

In the late 1930s, the Rural Electrification Administration strung electric wires to all the farms, and that white enamel washing machine with the labor-saving clothes wringer in the Bethesda store window moved right into the farm kitchen.

The market "has given us all a sense of security which even drought and depression cannot shake. We have helped our husbands pay off mortgages and back taxes. Our homes are now equipped with electricity, running water and other labor-saving devices. We have been able to give our children a higher type of education and our social life is brighter because we have cash in hand. Few of us would be willing to return to the old plan of accepting what our husbands can afford to give us," said Eleanor Waters.

Traditional dairy farming, once a reliable source of wealth, could no longer pay the bills. "Children were hungry, they had no clothes, no school. Now we are buying luxuries like tinned salmon, sardines, bits of sewing materials," said Waters.

By 1937, the 50-cent pieces that filled Margaret Davis's jar amounted to $750—more than $10,000 in today's dollars! She gave the money to John, who went to Rockville and bought a new Plymouth. Margaret herself never learned to drive and depended on her husband and sons to shuttle her to Bethesda.

At a time when the courthouse steps were lined with people waiting to see welfare workers and get relief, "farm women went in and saved the family farms," said the League of Women Voters.

Blanche Corwin's idea was bearing fruit in abundance.

The market not only "kept roofs over these women's heads and cattle in their barns, but (the farm women) have put many a child through school," said the Farm Journal.

The women expanded their purchasing skills and bought items for their homes. Furnaces, radios, electric mixers, hot water systems, and at least one electric chicken picker moved to the country.

They sold a wide variety of products. The Farm Journal reported, "You can buy firewood, plants, wild flowers, garden flowers, cocker spaniels, rabbits ... well, anything." Miss Nellie Hargett's colorful, mellow mints got a special plug, as FDR's housekeeper bought the sweets regularly. Mary Shorb's offering of unbaked dough to take home and bake sported a long line of waiting buyers.

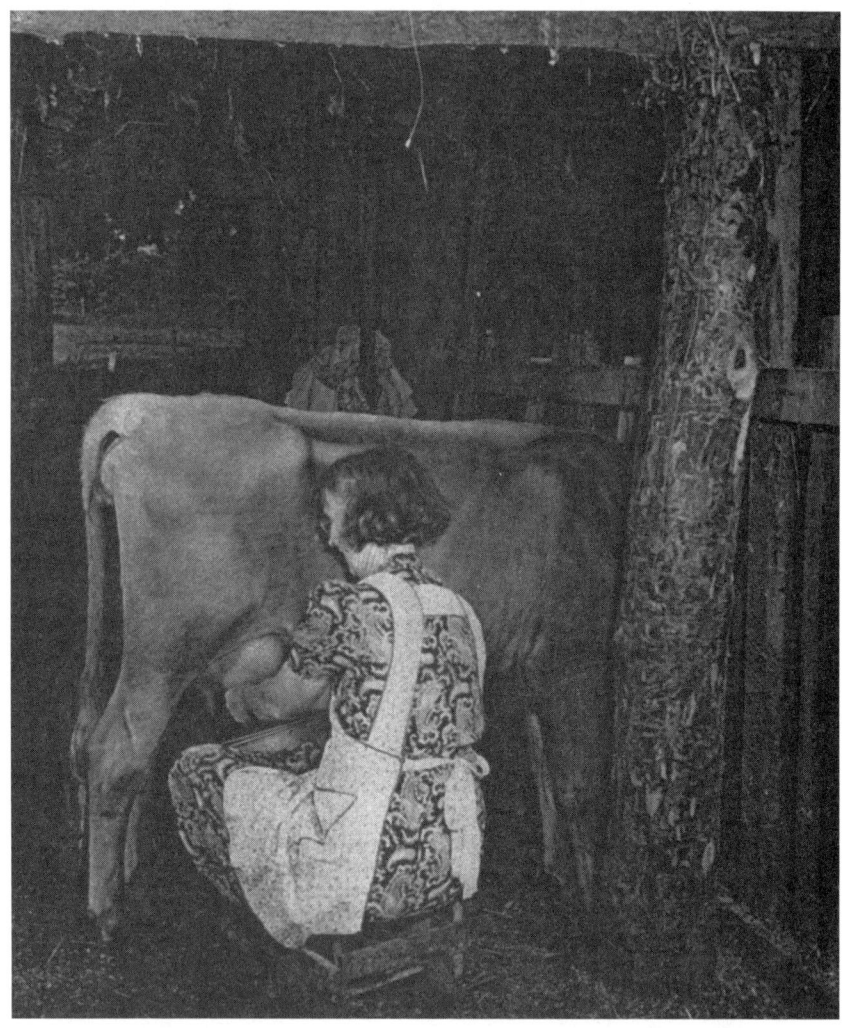

Unidentified but familiar, this American farm woman milks her cow. Tying her rickrack-trimmed apron on one side, she pulls an upside-down apple crate under her and grabs the milk bucket with her knees. A tree trunk holds up the roof. Courtesy of the Office of Communication, U.S. Department of Agriculture.

"You'd be surprised at the quantity of tempting food displayed in each small shop," said the Woman's Home Companion.

Reader's Digest climbed on the bandwagon. "The Montgomery County Farm Women's Cooperative Market at Bethesda, Maryland, is more than a shopping place for discriminating housewives from the District of Columbia . . . . It represents a method of beating the farm problem." Third in a series titled "Depression Born Businesses," the *Digest* agreed that the women had paid mortgages, financed college educations, and modernized farm homes. Prices were slightly higher than in the local grocery stores, but "Washington housewives cheerfully make the trip and pay the difference because of the high quality and because they find here specialties they can get nowhere else."

The *Digest* found new items: "[C]ornmeal mush packed into a loaf to be sliced and fried; beans baked in large deep pans and sold in cartons, hominy . . . homemade sausage and scrapple, candies and cookies; chickens beautifully dressed and stuffed . . . pots of herbs, bunches of cress, jams and pickles and flowering plants."

Mrs. John Darby sold 2,500 turkeys a year! Mrs. Edward West sold 60 dozen rolls a week; Mrs. Mary Hargett sold 2,000 jars of jelly "and over 60 gallons of chopped pickles a year."

Nannie Ray made pretty little terrariums she called Winter Gardens. She turned a glass gallon jar on its side and planted little wild things she found in the woods. Decades before anyone imagined that wildflowers might become extinct, Montgomery's streams were banked with spring beauty, Virginia bluebells, wild bleeding heart, wild orchis, wild ginger, Jack-in-the-pulpit, bloodroot, ginseng, turtlehead, cardinal flower, bluets, shepherd's purse, ferns of many varieties, mosses, and many more plants.

Nannie's terrariums graced Washington's winter homes and promised spring.

Another president noticed the market. By special order Miss Nellie Hargett made a large batch of her "locally famous" mints for President Truman's inaugural celebration.

On Monday, January 22, 1945, at their annual meeting, the members of the Montgomery County Farm Women's Cooperative Market burned their mortgage.

The Sentinel switched sides and became a firm supporter. "The only farm woman's cooperative market of its kind in the world," the newspaper proclaimed. No one disagreed, and the affirmation held.

Success crowned Charlotte Potter's cookies.

>
> **Valley Forge Bullets**
> 1 cup lard
> 1 teaspoon cinnamon
> 1 teaspoon baking soda
> 1 1/2 cups sugar
> 1 teaspoon salt
> 3 eggs
> 1/2 teaspoon cloves
> 1/2 cup molasses
> 2 cups raisins
> 3 cups flour
> 1/3 cup water

Pinch off pieces of dough. Roll between your hands. Flatten. With a paint brush, shine tops with 1 egg white, 1 tablespoon water.

>
> **Charlotte's Filled Cookies**
> Filling:
> 1 cup chopped raisins
> 2 tablespoons lemon juice

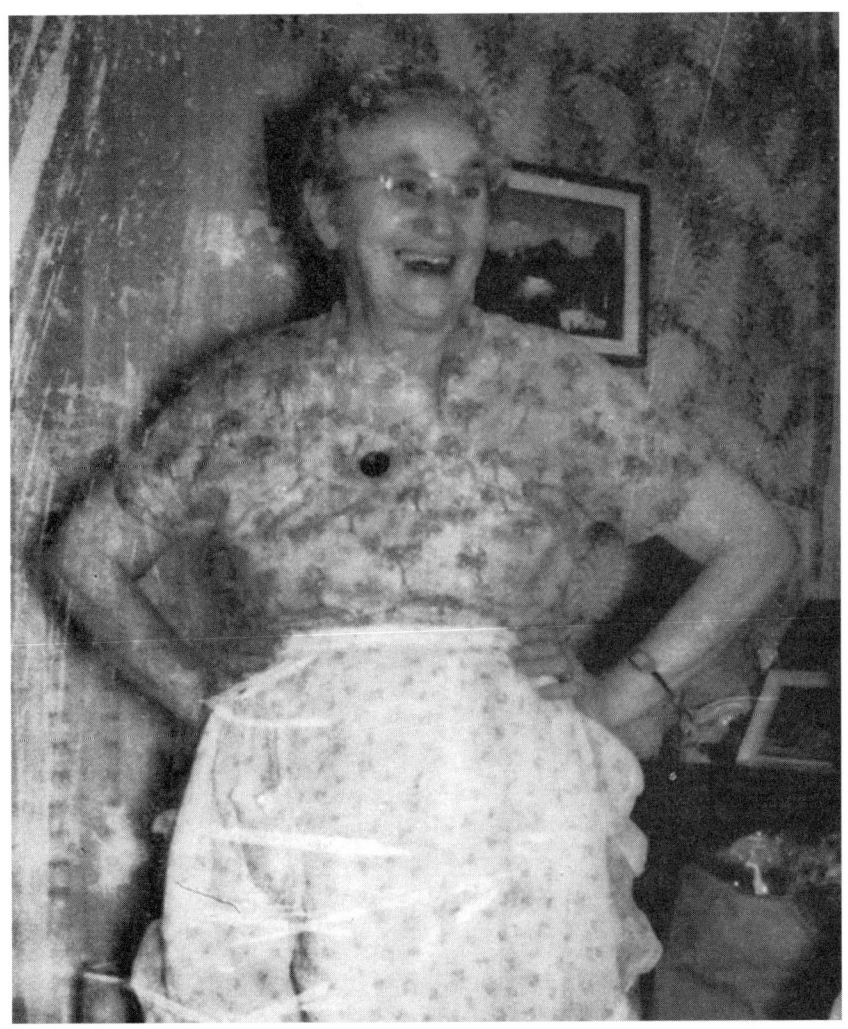

Molly Gladhill wears a frilly apron on Sundays. Courtesy of her daughter, Olive Gladhill Stup.

1/2 cup sugar
1/4 cup boiling water
1/8 teaspoon salt
Cook until thick.

Cookie dough:
1 cup sour cream
1 egg, beaten
1 cup sugar
4 cups flour

Roll dough and cut out rounds. Place one round on baking sheet. Add one teaspoon filling. Cover with second dough round. Pinch edges. Repeat with remaining filling and dough. Bake at 350° about 15 minutes.

Here's a popular recipe from Elsie White Daniel that sold well in the market:

Ice Box Rolls
1 cup boiling water
1 tablespoon shortening
1/4 cup sugar
1 yeast cake
3/4 teaspoon salt
1/4 cup lukewarm water
1 large egg (or 2 small ones), beaten
1 teaspoon sugar
4 cups sifted flour

Mix boiling water, 1/4 cup sugar, salt, and shortening. Cool. Soften yeast in lukewarm water and 1 teaspoon sugar. Stir in first mixture; add eggs and 2 cups flour. Beat well. Add rest of flour. Beat well. Cover with waxed paper and put in ice box overnight. Should double in bulk. Shape to suit, paint with melted butter 3 hours before baking. Bake in hot oven, 15 minutes. Yield: 2 dozen small or 20 large.

Chapter 5
Choices

The phenomenal success of the Montgomery County Farm Women's Cooperative Market saved farms and gave power to the women involved. All farm women who survived the 1930s were strong, smart, energetic, and courageous. They had to be. Although the market was magical for some of them, morphing despair into comfort, some Montgomery farm women found other ways to emerge intact from the Great Depression.

Betty Birgfeld King, Macie's sister-in-law, married Merhle King when she was eighteen years old and moved with her new husband into Brink, the King family farm. The year was 1926. Betty's new home was a historic manor house of three stories with wide porches, white gingerbread trim, and a big basement. A carved oak staircase swept upward to the third floor, and marble fireplaces warmed each room. It was quite a bit for a teenage girl to handle.

The home was lovely, but she and Merhle needed a way to earn income. They bought a small herd of dairy cows, which used all the cash the young couple had accumulated. The health inspector arrived. He tested each cow for tuberculosis and condemned every one.

Brother Jim King gave them two of his tuberculosis-free cows. Betty and Merhle borrowed enough money for a trip to Canada to buy a few more heifers. This was a time to strain their souls, but Betty and Merhle plunged into a strenuous work cycle that would have broken a lesser woman and man. Betty joined Homemakers. With advice from Blanche Corwin, she bought a few chicks, and Merhle built her a chicken house. The chickens multiplied and soon she had a good supply of eggs. She packed the eggs her family didn't need into a large, woven basket and sold them to her neighbor, Ada Main, who made cakes each week for her stall at the Farm Women's Market. Betty had money for her family and Ada had fresh farm eggs for her cakes!

Another sister-in-law of Macie's, Elizabeth Fulks King, lived on a farm between the towns of Gaithersburg and Rockville. She and her husband, Lawson King, kept the family financially stable through the depression by clever diversification and hard work. Every day King sent a fleet of trucks to collect milk from local farms, add the product of their own five farms, and deliver the milk to Thompson's dairy in Washington D.C. to be processed and distributed. Looking around at neighboring foreclosures, he and Elizabeth were probably scared spitless, fearing they, too, could become destitute.

Lawson was tireless, and took every opportunity to earn a little money here, a little money there. He milked the cows, hired drivers to forage the county collecting milk, hauled sod for Bill Wilmot, and "peddled everything everywhere he went," said James Wriley Jacobs, Lawson King's son-in-law.

There was a commodious, gracious house and a large dairy barn. Across the broad high roof the words THOMPSON DAIRY provided efficient advertising. Circling the barn were several houses for farm workers. Elizabeth's job was to feed the people.

"They ate very well," said Jacobs.

The milk-truck drivers came in around 8:00 or 9:00 p.m., and Elizabeth would run down to the basement and cook their dinner on her woodstove.

Elizabeth was always cooking, said her daughter, Betty Jean King Jacobs.

Good farm help was hard to find during World War II. When Lawson needed extra hands, he looked to the nearby German prisoner-of-war camp. Each morning very early, a busload of prisoners was driven through guarded gates, delivering experienced German farm workers throughout the county. The prisoners were willing and able. And free! Friendships developed. Although their wardens sent along a brown paper bag lunch, Elizabeth King cooked for them. She believed a sandwich was insufficient for a farm laborer. Each captured German was "some mother's boy," she told Betty Jean. But she made sure her daughter stayed inside the house when the prisoners were at their farm.

Lawson was a master at saving money. He purchased grocery items in bulk form, and taught his daughter his own personal slogan: "If you can't pay for it, you don't need it." He was not a borrowing man.

Under the name Irvington Farm, Lawson King built the world's largest Holstein herd and sold his champions on the international market.

Elizabeth and Lawson became important people in Montgomery County, counting among their personal friends the wealthy and powerful. Unannounced guests were a daily occurrence, and Elizabeth cooked and cooked, feeding bankers, politicians, wealthy investors, and her family, as well as the German prisoners.

King established a car agency on a corner of his dairy farm, and a farm supply store and tractor dealership in nearby Gaithersburg.

Like the Quakers of an earlier century, Lawson King's wife took care of his needs, and he could focus his energy on inventing ways to create a legacy. At the end of his day, he returned to Elizabeth and the warm clean home that smelled of fresh bread, roast beef, and apple pie. Lawson and Betty King survived the depression and became hugely successful.

Chapter 6
Changes: The 1950s to 1970s

Prior to the 1950s, there was a clear-cut dichotomy in the roles of men and women who earned their living on farms. There was men's work and there was women's work. Women crossed the line when an acute situation arose, earning the family income during the Great Depression. But it was unusual for men to assume the role delegated to women. Men worked long, hard hours in their fields and barns, often becoming prosperous, and women cooked, cleaned, and tended the kids. During the 1950s and '60s, however, the traditional roles began to disintegrate.

"In the early 1960s, farming changed," said Cecile King Jones. The residual effect of World War II was partially blamed for the upheaval, she said. Mrs. Jones's grandfather, Elias Dorsey King, paid day labor $2 a day, considered a fair wage before World War II. During the war, Elias King watched his farm workers sail away to Normandy, but King had the help of German prisoners of war. When the war was over, the Germans went home and the local farmhands came back—but not to work for $2 a day. Good farm workers could easily get a job at the new National Institutes of Health in Bethesda for a regular paycheck plus paid vacations and health insurance. Compared to that, $2 a day seemed mighty slim.

The extension service began to involve more girls in 4-H clubs. The second picture of women in the Montgomery

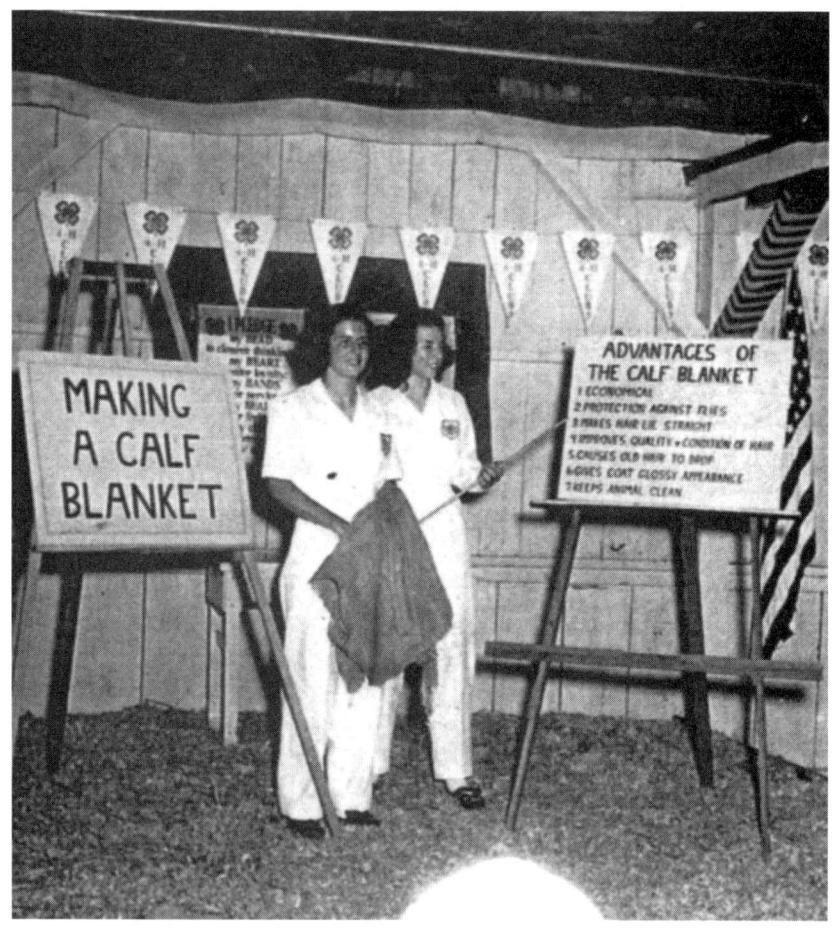

Barbara Ann and Joyce Riggs with their winning 4-H Project, circa 1948. Extension Service archives.

County extension agent's annual reports shows two sisters, Barbara Ann and Joyce Riggs, demonstrating "Making a Calf Blanket." On the same page is a picture of the boys' 4-H club getting ready to go to Europe. Roles hadn't changed much; farm girls made calf blankets and farm boys went to Europe. But change was in the air, and both Riggs sisters graduated from college, Joyce in economics and Barbara in agriculture.

Barbara Riggs sailed right through college, got top grades, and applied to graduate school. But America wasn't ready. Few women were invited into the top graduate programs in science. Her major professor told Barbara that if she were a man, every college with a graduate school in agriculture would have been at her door. As it was, eight graduate schools recruited her. On the cusp of social change, Barbara chose to set graduate school aside, and married farmer Stanley Stiles.

Turf farms were sprouting, and Frances Wilmot reappeared on the farm scene. In 1936 Frances Wilmot's parents had decided that owning a farm would be delightful, and that perhaps they would again see their horse-crazy daughter, who could not stay with Macie King forever. They bought Summit Hall in Gaithersburg at auction for $13,500, including the big, historic house, several outbuildings, and 135 acres. Because Mr. Wilmot worked for the government and got a regular paycheck, he did not need the extra income, so he rented his land to Lawson King.

Frances was ecstatic. She purchased the beloved mare Lady from Macie King and rode her to Summit Hall, a distance of about ten miles. Frances was then fifteen and thoroughly at ease with Lady. The two of them forded streams and maneuvered the dirt roads of rural Montgomery, meeting very few cars.

Summit Hall was quiet and comfortable, but in the early 1950s the farm changed direction, thanks to Frances's brother

Bill, and began a sprint to national fame and considerable fortune. Bill came home from World War II and decided to try farming. At first he sold turf cut from unimproved native grasses. Wilmot called his product "pasture sod," but he was not satisfied with his crop. Seeking improvement, he turned to the U.S. Department of Agriculture (USDA) in Beltsville, Maryland. Try cultivating the sod, he was told. Bill Wilmot did. He cultivated and fertilized his soil, purchased good grass seed, and sold his improved turf sod. Frances handled the office work in the former tenant farmers' house, where a glassed-in office had been added.

Then the scientists at USDA urged Wilmot to try their newly developed grass, zoysia. Bill tried the new cultivar. The grass looked good, but harvesting the plugs was labor intensive, requiring each 1-inch group of tender roots to be hand-lifted from the ground. A cousin visited, saw the situation, and invented a contraption with an engine that popped out the plugs for easy pickup.

Fortune smiled on the Wilmots. Arthur Godfrey, who owned a farm in Virginia, bought a yard full of Summit Hall sod and was so pleased, he devoted a segment of his weekly radio program to a discussion of the Wilmots' crop. The orders flowed in. Even the White House placed an order! Frances answered the phone and juggled the paperwork.

Frances and Bill hired Donna Oden and five women to pack the plugs for mailing. They stood on the cool concrete floor of the former granary every day, "late spring til fall," making up the orders. Fifty plugs to a box were packed in wax-coated cardboard crates "like egg cartons," said Frances.

Donna wrapped each package in waxed paper to prevent moisture from ruining the mail, and small orders were carried to the post office. Orders of ten or more double boxes (1,000 plugs each) were sent by rail, which was even closer to Summit Hall than was the post office.

"Growing sod," said Frances, "was economically successful."

Following the end of World War II, the housing market exploded. Houses replaced farmland as Montgomery County began turning into a city. New families did not like to smell manure, and each new house wanted grass. Summit Hall Turf Farm was a beneficiary of this shift in land use, but local farm businesses began to falter. In Germantown, a medical practice bought a house next door to a country butchering business, and suddenly abattoirs were outlawed within the county, forcing farmers to drive long distances to market their livestock. Day after day farmers saw the smiling face of a friendly developer. Selling their farms seemed like a good idea.

In the upper reaches of the county, farming was still an important and prosperous business. Peggy Johnson of Dickerson remembers the fall festival surrounding a country butchering. All farmers in the neighborhood came to the event. Men worked with the pig; women worked in the house preparing a huge meal. One autumn day, the Linthicums of Seneca Ayr Farm held a butchering. The men had completed their chores except for the pudding, a treat made with all the leftover scraps of meat plus cornmeal to thicken. Someone had to stir the iron caldron of pudding, and Arthur Johnson, Peggy's young husband, was chosen to do the job. Ethel Linthicum asked him what she should hold back for his dinner. Arthur said, "Miss Ethel, just save me a piece of your apple pie."

Mrs. Linthicum did, and when the men had finished their meals and gone back outside, Arthur came in to eat. "He ate the whole pie—with vanilla ice cream on top," said Peggy.

As cities expanded, new roads were built and local governments needed more land. Farmers who refused to sell could be faced with laws of eminent domain and lose their options, or they could negotiate a price and sell. But they could not continue traditional farming.

Lawson King's neighbor said new road construction cut off his cropland and pastures from the rest of his farm and he could not get his hay into his barn nor his cows into the dairy. The world turned, and farmers, once the heart and soul of Montgomery, found themselves losing their lands, their hired help, and their position at the top of the county's economy.

As farmers watched and worried, the Atomic Energy Commission, National Bureau of Standards, and IBM took over vast areas of farmland.

Farm workers swarmed to the new facilities, and overnight more farmsteads fell to housing developments. In Wheaton, Maryland, nearly 47,000 houses were built in the 1950s to 1960s and a new type of shopping facility appeared: an entire arena of city shops under one roof with potted palms in the center. Delighted consumers flocked to the glittery mall, and traditional stores failed in towns and in the country.

The average price of one acre of farmland in Montgomery County soared from $301 in 1954 to over $4,000 in 1964.

Alert county commissioners appointed a committee to study land use and recommend a master plan. From 1950 to 1957, these five men talked with farmers and developers and studied the county map. In announcing their findings, the Upper Montgomery County Planning Commission recommended that one-third of the county, approximately 100,000 acres, be set aside for agriculture and open space.

The lives of two women best illustrate the changing farm life during the 1950s.

A traditional farmer she was not, but Ruth Elizabeth McCormick Tankersley, born in 1923, created an agricultural niche for herself and made the Arabian horse one of America's favorite breeds. Her close friends called her Bazy; to the farm staff, she was Mrs. T.

Bazy was born to power. Her grandfather, Senator Mark Hanna, was dubbed "President-Maker"; her mother was U.S. Representative Ruth Hanna McCormick, and her father was Senator Medill McCormick. Her stepfather was Congressman Albert Simms. Also on the family tree was Bazy's uncle, Robert McCormick, publisher of the *Chicago Tribune*.

When Robert McCormick bought the largest newspaper in Washington, D.C., the *Times-Herald*, he appointed his niece as publisher, the first woman to hold such a position. Suddenly, Bazy was catapulted into the center of power and politics, and her every action was open to public scrutiny. For nearly two years, 1949–1951, Bazy was arguably the most influential woman in the nation, publisher of the capital city's largest newspaper. *Time* magazine announced the appointment, anointing Bazy with the title "crown princess," under the caption, "A Castle for the Princess."

All too soon the castle crumbled. Uncle McCormick gave her the job, and just as easily he snatched it away again. *Time* had called her "a plain, unexcitable grey-eyed blonde" in 1949, but nineteen months later, the magazine gave Bazy a quite different personality:

"High up in Chicago's Tribune Tower, the door to Colonel Robert R. McCormick's sanctum flew open. Out strode the colonel's niece.... Mad as a wet hen, she took the elevator to the lobby, hustled off to her suite in the Ambassador East Hotel." Bazy was fired after "a heated showdown" with her Uncle Bertie.

Bazy and her two children retired to Al-Marah, the Arabian name she gave to her 50-acre farm in Bethesda. Divorced, she married Garvin Tankersley, whose job as assistant editor

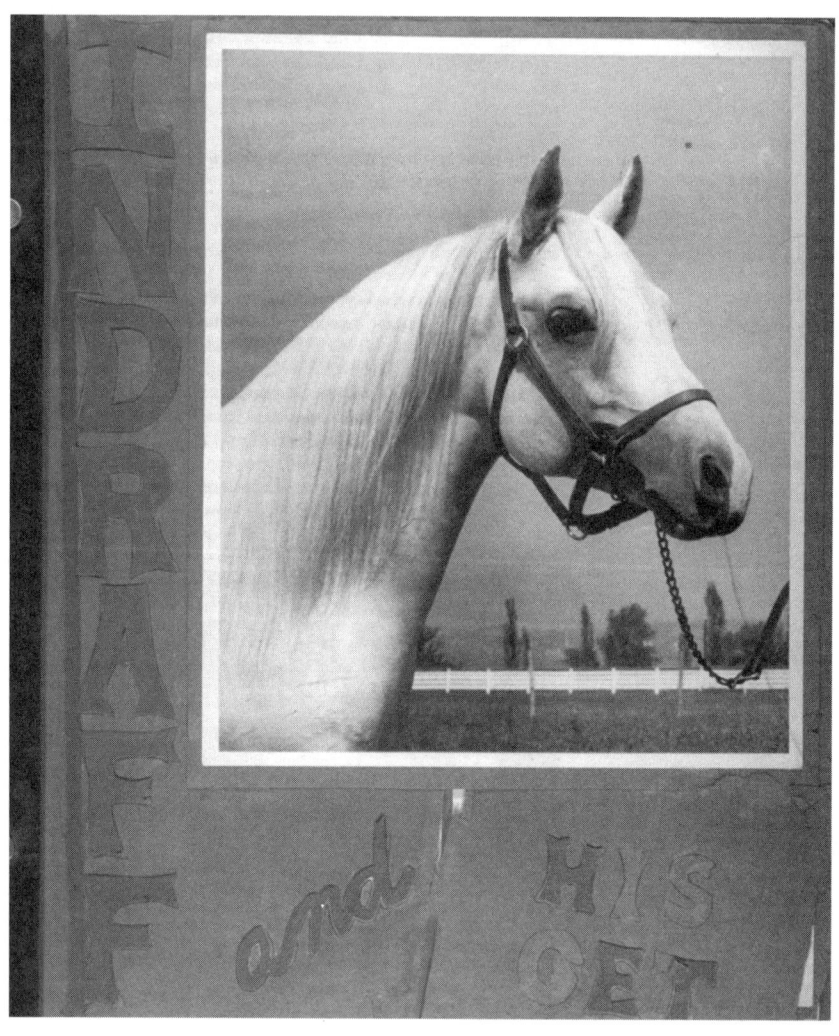

The magnificent stallion, Indraff, of Al-Marah. Courtesy of Ruth McCormick Tankersley.

at the *Times-Herald* had been similarly terminated. Now there were four children at Al-Marah: Garvin Tankersley Jr., 11, and Anne Tankersley, 9, who spent their summers with their father and stepmother; and Bazy's children, Kristie Miller, 6, and Mark Miller, 4. Bazy was thirty years old.

Being financially independent, she could afford to take chances. Bazy chose farming. Her lifelong passion was the Arabian horse, and she plunged into expanding her herd and breeding the magnificent equines. Breeding stock was imported from England—not as single individuals, but as groups of blooded Arabians—and Al-Marah became known as "the largest breeding farm in the United States."

Bazy and Tank purchased an additional 1,500 acres near Barnesville, Maryland, and added Hereford and Santa Gertrudis cattle to their agricultural pursuits.

Bazy could breed them, raise them, train them, but could she market her livestock? Individual sales hovered at $1,500, with the majority selling for $2,500 to $10,000. Indraff, a magnificent, silvery stallion, had produced gorgeous offspring, and Bazy was determined to get the best prices. In 1954, searching for a premium market, Bazy held the "first major public Arabian sale ever," and averaged $1,625 for each of 40 animals. Two years later the second Al-Marah auction brought in $100,776 for 48 horses, an average of $2,099.50 each!

Another child joined the Tankersley household: Joanna Sturm, the granddaughter of Bazy's mother's friend, Alice Roosevelt Longworth. Orphaned at age ten, Joanna fit right in at Al-Marah, cheerfully participating in equestrian activities.

Summers were a childhood dream come true. After hours of riding, the Al-Marah kids dismounted at the low-lying barn, groomed their mounts, and put them away. Walking up that last hill to the house, one foot pushing the other, growing hotter and hotter, sweatier and sweatier, sun beating down, horse hairs scratching their necks, the children thought they would never make it. But once at the top, there was the pool! A cool refreshing swim was followed by lunch, and then by games with Bazy. Bazy's kids loved this time together. They called their favorite card game Pounce.

"Bazy loved to laugh," said her stepdaughter, Anne Tankersley Sturm. "Still does!"

Interested always in education, Bazy joined with others to introduce horse management courses at Al-Marah. They hired the nation's top trainers and veterinarians. Horse owners from across the nation came and explored topics pertaining to horse nutrition, training, first aid, housing, and more. Before long the courses were offered in Illinois, Texas, Michigan, Colorado, Indiana, New York, and Tennessee, as well as at Maryland's Al-Marah.

When the Washington beltway was constructed, Al-Marah became part of the city inside the concrete loop. Bazy and Tank bought another 2,300 acres near Barnesville, and sold the Bethesda farm. They built a large house and an even larger barn and covered arena. At a crossroad near her house, Bazy's group constructed a museum devoted to the Arabian horse.

Friendly and interested in everyone, Bazy focused on mixing people from all walks of life. "She never wanted to isolate herself," said Anne Sturm.

Bazy designed a plan for developing a portion of her farm. It would be a horse owner's neighborhood, a small community of individually built houses, each with a minimum of ten acres, facing a winding road. A historic farmhouse and bank barn were already on the property, and these became a clubhouse and barn for the small community. Bazy named her pretty hamlet "Conoy."

In 1967, at age 46, Bazy became a mother again, welcoming Tiffany Tankersley. But Bazy was not about to slow down.

She began to think of installing a primary school in part of the Arabian Museum.

Barnesville, settled in 1747, is a happy place of peaceful contentment and friendly folks. Gentle farming is the established industry; sidewalks keep the children safe, wisteria climbs the trellis by the old town well, and every Sunday the perfumed air echoes with the melodious bells of St. Mary's Catholic Church and Shrine. Baptists gather, too, at their Currier and Ives setting, a white clapboard chapel with a slender steeple pointing to the sky.

There are plenty of barns in Barnesville, but no place to buy a loaf of bread, and residents work hard to keep it that way.

When Bazy Tankersley moved to town, Barnesville was splashed with glory. The Lone Ranger came to visit! Clayton Moore ("A fiery horse with the speed of light, a cloud of dust, and a hearty Hi Yo, Silver! Away") arrived to borrow a replacement for Silver from Al-Marah. Roy Rogers' Trigger also came to town, along with his trainer, and Walter Farley set aside his Black Stallion series to buy an Al-Marah Arabian.

Bazy produced a pageant to benefit the American Heart Association. On the day of the pageant, she dressed the family children and farm staff in bathrobes and turbans, slung colorful blankets on the Arabians, matched mountees with mounts, and sent them out to the hills surrounding the Conoy clubhouse. All the tickets had been sold, and the country roads began to fill with cars going to the performance. As the sun went down, onlookers could hear the mares and foals whinnying inside the barn, echoing excitement.

Off in the distance, the riders were glimpsed in the rolling farmland. Just as they crested the tallest hill, the moon rose behind them. Each frilled blanket and wound turban morphed into a perfect silhouette outlined against the bright glow of a full moon. The crowd held its collective breath.

At this instant, Bazy released the mares and foals, who sprang into the ring between the audience and the barn. The horses began to circle 'round and 'round, running, kicking up their heels, stimulated by the smells and sounds of many people. As soon as the horses were let out, Bazy pushed a button and the theme music from *Born Free* wafted gently over mares and foals and people.

Barnesville was never the same.

Bazy Tankersley became a legend in her own time, always generously supportive of a host of charitable organizations. When the Smithsonian National Zoo in Washington, D.C., needed to remodel quarters, the staff asked Bazy to provide a temporary home for wildebeests and two herds of zebras.

In the mid-1970s, Bazy, Tank, and Tiffany moved from Maryland to the southwest, where Bazy bought the Quarter Circle Double X Ranch, near Williams, Arizona, home of former congresswoman Isabella Greenway. She continued horse farming until, by 2001, Bazy had bred "at least 2,200 horses."

Not only a farmer and breeder of fine horses, Bazy is known for her philanthropy and her lifelong interest in education. The Ruth McCormick Tankersley Charitable Trust is one of the top forty grant-giving foundations in Arizona. A former member of the Board of Maryland State Teachers Colleges, Bazy has established three schools: Primary Day School in Bethesda, Maryland; Barnesville School; and St. Gregory's College Preparatory School in Tucson, Arizona.

A woman of great courage and commitment, the major legacy of Bazy Tankersley is the Arabian horse of America. Born to power and privilege, she could have led a life of leisure. Instead, Ruth Elizabeth "Bazy" McCormick Tankersley chose an equestrian farm.

Another farm woman of the same period but dissimilar circumstances made different choices.

Born in 1922, just one year before Bazy Tankersley, Elizabeth Beall Banks spent her early childhood on a farm in Howard County, Maryland. She walked behind a horse-drawn plow, kept chickens, and attended a one-room school. With her family she moved to Belward, a 300-acre farm in Montgomery County owned by her maternal grandparents. Here "Liz" Banks milked Brown Swiss cows and grew to love the farm. After graduating from the University of Maryland, Liz went to work teaching high school English in Rockville. She taught school for fifty years, coming home every day to change into overalls, drive the tractor, and feed the cows.

By the early 1960s, her parents had died and her siblings had left the area. Liz paid the inheritance tax and took over management of Belward.

She decided to switch from dairy to beef cattle and went to her neighbor, Mr. Herman Rabbitt. Would he sell her twenty of his nice Black Angus heifers?

He agreed. When they were delivered, Liz asked what she owed.

"Nothing," said Rabbitt.

Legend has it that Mr. Herman Rabbitt buried a fortune in his backyard and the money has never been found. Liz Banks told her friend and veterinarian, Dr. Susan Moxley, this tale about the late entrepreneur:

One day Mr. Rabbitt went to the Chicago Stockyards. Dressed in his usual overalls, with manure on his boots, he liked what he saw and began buying. Soon the local bank got a call from Chicago saying they had a man up there looking like he hadn't two cents to rub together who was buying everything in sight.

What was his name? asked the bank official.

Herman Rabbitt.

Oh, said the banker. If Mr. Rabbitt writes you a check on the back of a paper bag, cash it. If he wanted to, Mr. Herman Rabbitt could buy the entire stockyard.

Back at Belward, Liz Banks improved her fences and began to raise Black Angus.

At school, her students called her "Battle Axe Banks," according to Dr. Moxley, "but she was kind and gentle with her animals. A great cattle woman." Disruptive pupils were assigned detention on her farm, and several of them returned every year of their own volition, learning about farming and providing needed help.

Mr. Rabbitt gave Banks a recipe for feed: ground ear corn and chopped hay with molasses poured on top. He called it gingerbread, and the feed made Liz Banks's Black Angus cattle thrive. The diet was expensive and time-consuming, but nothing was too good for Liz's steers.

As the city of Rockville grew outward, Banks's troubles began. Woodland was replaced by houses, displacing the groundhog population, and Belward became overrun with the furry, voracious, hole-digging creatures. Liz hired a man to shoot a few. One shot, and a policeman appeared. Liz learned that a gun could not be fired where she lived, and she paid a $5,000 fine.

Geese flew in and nearly denuded her pastures. Liz bought a Screamer, a harmless device that makes a loud, rifle-like noise, encouraging geese to go elsewhere. What a sight it must have been to see eighty-year-old Liz Banks standing in her field firing the noisemaker.

Liz had a warm spot in her heart for Johns Hopkins University. During her mother's final illness, the elderly woman had been treated kindly at Hopkins's Baltimore hospital. In 1989, believing Hopkins would attend to her wishes, Liz essentially donated Belward to the university, accepting $5 million for 138 acres with an estimated market value of $40 million.

Although Liz hugged her trees and placed her thin body in front of behemoth bulldozers, she was not able to preserve her farm. Even before the sale, part of her land was taken by eminent domain, a new highway built, and roads widened into her pastures. Her agreement with Hopkins held that a small portion of the land could be developed at the time of sale and the remainder subdivided after her death, but she would keep her farm with its eighty Black Angus cattle until she died. She believed the agreement included leaving her woodland untouched, but every huge old oak was removed. Blasting went on for days as an enormous research facility sprang up on thirty acres of Belward. Heartbroken, Liz did not hesitate to express her fury. Hopkins planted saplings.

At the age of eighty-three years, farmer Liz Banks, no longer able to shoot the Screamer or even to stand without the help of her walker, lay down on her old bed in her dilapidated farmhouse and quit the battle.

At her funeral, Merle Steiner spoke about his old friend's feelings toward her Belward. "She cried over not being able to preserve it as a farm. . . . She wanted to continue the heritage of the farm."

The *Washington Post* announced her death with an article headlined, "Farmer's Death Lifts Restrictions on Property."

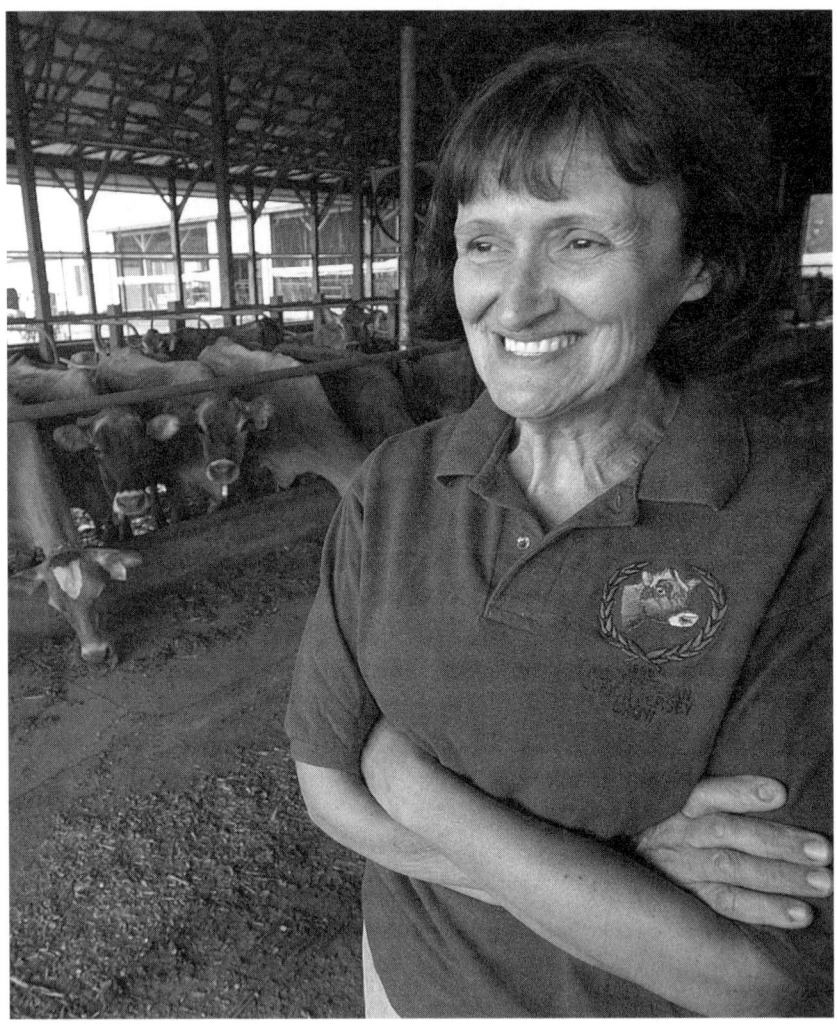

Janet Stiles, Dairy Woman of the Year 2006, with her 100 milkers. Courtesy of the *Frederick News-Post*, Frederick, MD.

Chapter 7
Farm Women of the Twenty-first Century: Making Money

Farm women today face the same financial challenge as in previous generations, but make use of technologies undreamed of by their predecessors. A prime example of a contemporary farmer is Janet Stiles, dairyman.

The Shenandoah River is narrow and fast-moving as it rushes through the town of Boonsboro, Maryland, founded in 1792. Opposite an abandoned mill, a dam gentles the river until it drops, splashing huge gray boulders in a shimmering waterfall. Overlooking this breathtaking site is Shenandoah Jerseys, the farm/business/home of Janet Stiles.

In October 2006, Stiles was named Dairy Woman of the Year by the World Dairy Exposition in Madison, Wisconsin. Her thin frame looks deceivingly fragile, but her large, muscular hands reveal the truth. Each day begins earlier than the farm women of yesteryear. Stiles rises at 2:30 every morning to check pregnant cows, care for new ones born during her short night, shovel alleyways, spread fresh bedding, and examine the health of her Jerseys. Once she is content that all is well, she sets to milking one hundred cows. All of this she does by herself. Although she has staff for the countless chores, Stiles does all of the milking herself. In late afternoon, the process is repeated, every day, day in and day out, Christmas or the 4th of July—up at 2:30 a.m., to milk those cows.

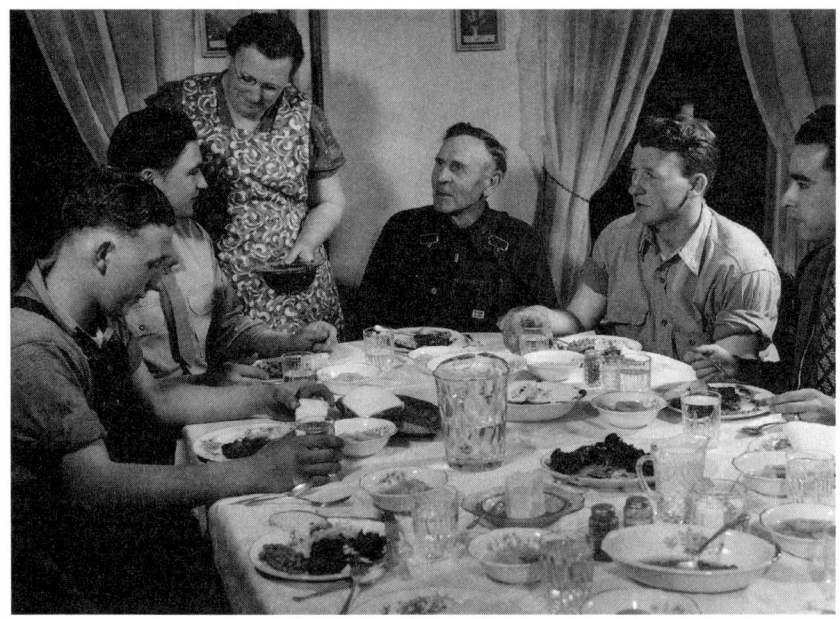

Mrs. Stiles wears her starched and ironed feed-sack apron as she serves her husband, sons, and a visiting soldier; circa 1943. Courtesy of Library of Congress, Prints & Photographs.

Milk is the major product at Shenandoah Jerseys. Each cow produces 60 to 70 pounds or slightly less than 4 gallons of milk a day. Technological advances have produced Stiles' state-of-the-art milking equipment, rushing milk from the cow's udder to a stainless steel refrigerated tank with outside access for the milk truck that picks up every other day.

Stiles' cows are an elitist club; it's hard to get in, but once a member, happiness comes automatically. Their alert, pink ears and soft, forest-brown eyes exude peaceful contentment. Shenandoah Jerseys are individually selected by Stiles, a well-known judge of dairy cows. From the time she was a child growing up on a dairy farm and continuing through 4-H, the University of Maryland, years as an extension agent, and judging dairy cows from East Coast to West and beyond to Australia, Janet Stiles has honed her judging skills to a fine edge. Her own ability informs her of the best cows to purchase and the best breeding sires for her herd.

"The correct cow . . . makes the difference between eking by or in the gravy," said Stiles.

All her breeding is done by artificial insemination. Reams of paper are scanned each year, choosing the perfect choice of mate for each cow.

"If she is a little weak in the legs, I choose a bull that has produced offspring with strong legs. Or if she is short on milk, a bull whose calves are heavy milkers," said Stiles.

Selective breeding through many generations has produced a superb herd of pedigreed jerseys. With orders from buyers wanting bulls as well as heifers, Stiles can sell any calf her herd produces.

With the recent construction of a commodious dairy barn, Stiles' dairy operation is all under one roof. Her office is in one wing and her milkers live in a large, airy, high-ceilinged portion of the main barn. Each cow is named and is, Stiles claims, spoiled rotten. Nothing restrains their heads as they contentedly munch nutritious grain and hay. Two months before her calf is due, the expectant cow is relieved of duty and given a vacation in acres of rolling, green pasture overlooking the river, with nothing to do but eat green grass and enjoy the view—a real cow paradise.

Janet Stiles has no transgenic cows—animals that are bred with DNA that is specifically designed to cure disease. But if transgenics could have saved her husband of twenty years, Tracy Stiles, from an early death in 2000, she would have switched to transgenics "in a minute," she said.

There is a "real crisis right now" in the dairy business, said Stiles, an active board member of the Maryland-Virginia Milk Producers Cooperative. Prices paid to the farmer are "twenty to thirty percent lower than last year," she said, adding, "the only way you can make real money farming is to sell your land."

Stiles is not selling her land. She is a milk evangelist, promoting and educating, singing the benefits of milk at every opportunity. An articulate proponent of her product, she has convinced the high schools in Washington County, Maryland, to include flavored milk in the vending machines used by students. As she sees it, milk offers much greater nutritional value than do sodas; in fact, the two are far apart. Milk is good for kids. Flavored milk tastes delicious, and students are responding with enthusiastic acceptance. As part of a wellness initiative, Washington County has placed a milk vending machine in every high school at a cost of $4,000 each, half paid by the dairy cooperative.

Local McDonald's, Wendy's, and Giant grocery stores now sell flavored milk. "Kids like milk and drink milk," Stiles said.

From a small flock of sheep to a large herd of llamas took twenty years of effort, but Elena Stamberg of Graceful Legend Llamas has one of the most successful llama herds in America and sells her breeding animals for top dollar.

"It is possible to take in $75,000 a year from the sale of three llamas," she said.

Stamberg bought Cornelius, her first llama, in 1987. Her children's 4-H project was sheep, but Cornelius made clear his superior intelligence and hardiness.

"Llamas don't have the health problems that sheep have," said Stamberg. "Unlike sheep, when llamas are injured or sick, they let you know and readily accept treatment." Llamas are naturally curious. On a recent day, students at the country day school next door were walking along the fence, looking at the llamas. In a group, all the llamas ran down to the fence and looked back at the children!

"When a cria [a baby llama] is born," said Stamberg, "the entire flock gathers 'round in a circle. They sniff the baby. Llamas never injure the little ones, but step over them and patiently push them away when one mistakes the wrong udder and tries to nurse someone else's mom."

Rain, snow, or blazing heat, the llamas, like all farm animals, must be fed and kept clean. Each individual has a separate feed pan and is allotted the exact amount of grain and vitamins necessary for individual needs. Hay is carried from the hay shed, fed in specially designed containers high off the ground in various places throughout the llama barn, and fresh water supplied. Llama manure is a favorite with gardeners as it is encapsulated in a biosoluble membrane and has little odor. The substance releases nutrients gradually and is an excellent natural fertilizer.

Graceful Legend Llamas travel to shows from California to Maryland, and come home with championship ribbons. A superb judge of quality breeding stock, Stamberg purchases only the best and has an excellent line of show stock.

Her Suri llamas grow lustrous fiber, which Stamberg uses to create art, selling her fine, hand-knit scarves for $250 and her unique sweaters for nearly $1,000.

"I never make the same thing twice," Stamberg said, holding up a fantastic shawl of many knitting patterns and subtly blended shades of llama colors—black and white, with shades of grays and browns.

Hard work and a dynamic, focused spirit have made Elena Stamberg a successful farmer of the twenty-first century.

"Diversify," said Jane Tabb of Kearneysville, West Virginia.

Tabb grew up within a few miles of Washington, D.C., but sought work on farms, graduating from Virginia Tech with a degree in dairy science. She accepted a job milking cows and married the farmer.

Farming has good days and bad days, and you learn to roll with the punches. When Jane and her family could not get labor to help milk their 300 Ayrshire dairy cows, and then the price paid for milk dropped dramatically, they sold the herd. Today they own 400 Angus beef cattle, grow straw, and sell composted manure and mulch.

Tabb speaks for farmers. She is involved with local politics. She was elected to the Jefferson County Commission for five years and was the commission president for one year. She started the annual Jefferson County Farm Tour and participates in farming activities throughout the state.

Diversification has paid off for Jane Tabb's family farm. It's a good life without the large dairy herd. But milking cows got into the blood of this former city girl, and she keeps a few Ayrshires to show at the county fair.

Diane Savage Geary, now a senior staff member of the Montgomery County Extension Service, grew up on a dairy farm. She cooked meals for her father, brothers, and hired help, and picked up replacement parts for farm machinery. Chickens were her responsibility, and she tended them, butchered them, fried them, and laid them on a platter, which she set before her hungry menfolk. Food was her special realm. The Savage family had their own beef, pork, and chicken, a garden full of vegetables, and fruit trees. What they didn't eat fresh was preserved.

Geary has fond memories, but fears for the future of the family farm. Seated beneath a wall of quilt squares and a tray depicting farm animals, Geary sighed and predicted that dairy farms are "dying out."

"Farms today are high-tech," she said. Genetics and scientifically precise feeding have revolutionized the dairy business. One cow can produce enormous quantities of milk and may bring a sale price of $30,000. With the sale of such a top pedigreed animal goes a contract for her embryos, so much for her bull calves, so much for heifers. Young farmers cannot compete for such an expensive cow, and it costs the same to feed and care for a lower-priced cow. High tech and low tech both are practiced on dairy farms today. Cows are fed and milked on schedule, barns are cleaned, and government regulations are scrupulously followed, translating into 12- to 14-hour workdays with no time off.

The Savage Farm tried added-value farming. They hosted farm tours for schoolchildren until they were forced to quit when one day a child, in spite of being told not to, picked up a cat and was scratched. All sorts of fur flew, the farm's insurance rates shot up, and the Savage family canceled all tours.

According to Diane, farmers are moving out of Montgomery County, but no farmers are moving in. Farmers work 365 days a year, some chores need to be done twice a day, and farmland is dear. In addition, farming in general has become very expensive.

"You have to have a huge bankroll to get into farming today," Diane says. "Unless a farm is inherited, along with several hundred thousand dollars of essential equipment, a young farmer will fail no matter how smart he is."

Epilogue

Women farmers still work hard, but enjoy it more. They raise horses, sheep, goats, and cabbages. The Montgomery County government helps by sponsoring four farmer's markets during the growing season. Farmers sell berries, sweet corn, flowers, herbs, and a wide variety of vegetables. Health regulations now preclude dairy products and home-baked goods.

Looking back through the years, it's clear that the Quakers of Sandy Spring left a legacy that will likely continue as long as a single farm remains in Montgomery County—the historic Farmers Club. Members continue to meet, share a meal, and discuss issues that are relevant to agriculture today. Men only. Women may be invited as guests, but may not become members. The Women's Mutual Improvement Association continues to meet, too. Women only.

Considerably changed from yesteryear is the University of Maryland Extension Service. No more Rural Women's Retreats, no more markets staffed by women only, no more home economics, and no more home demonstration agents. What was once referred to as "extension agent" is now called "educator." Former home demonstration agents are family consumer science educators. One educator holds classes in financial management and estate planning for farmers; another teaches integrated pest management; and yet another describes the proper use of weed control. Youth educators staff the 4-H clubs; one of them teaches girls to sew and holds a contest each year wherein outfits made with the girls' own hands are judged and awarded prizes.

The horticulture educator faces a situation that was unknown to Wardney C. Snarr. He needs a translator because most of his students speak Spanish only.

Douglas Tregoning is now the Montgomery County extension director and agent of Agriculture & Natural Resources. Hired in 1980, before the new titles went into effect, Tregoning is there to help the farmer. If a farmer has a problem, Tregoning gets right out to the field, diagnoses the cause, and prescribes treatment. He's good at what he does. Doug's great-grandmother, Molly Gladhill, was an early and longtime vendor at the Farm Women's Cooperative Market. Farming is in Doug's genes.

Tregoning is enthusiastic about agriculture. His daily contacts inform him that farmers are doing better on environmental issues than ever before. They use less-toxic pesticides, and the herbicide now being spread is much more benign than in previous years.

"Prudent managers" is the term Tregoning uses to describe his clients. The water quality is better; the soils are less eroded and contain more nutrients now than in earlier years, he said.

Harking back to the days of Blanche Corwin, an agent's job now has more to do with politics than with raising chickens.

"Today there are a lot more issues," said Tregoning. He struggles to find answers for questions such as, What practices should be permitted in the agricultural preserve? Should innovative septic systems be allowed? Transgenic wheat? Should farms bordering rivers be permitted to pump their own water, and possibly change their crop to townhouses instead of corn?

And then there is the problem of balancing two ways of looking at farmland preservation.

"People moving to the country think everything should stay the same and tell the farmer, 'I'll help you keep it that way,'" said the extension agent. "That doesn't sit well with the farmer, who is used to farming the land as [she or] he sees fit."

Different, too, is the Montgomery County Farm Women's Cooperative Market. The city has grown up around the inconspicuous, one-story structure. Multistoried office buildings replaced the abandoned storefronts seen by Blanche Corwin, and the streets of Bethesda are teeming with shops of every sort and with people from all over the world. The market is still open Wednesdays and Saturdays, and pleasant aromas waft from inside.

The Farm Women's Cooperative Market continues to feature friendly vendors and attractive merchandise. The outside yard, however, has changed. There has been a flea market in front of the building since the 1970s. No window-box geraniums. No white dresses and caps. During the 1970s, the county wanted to knock down the market and build more multistoried office buildings. Neal Potter and Bruce Adams, powerful local politicians, fought successfully for retaining the market and controlling the escalated taxes. Ownership today remains with the stockholders. Men as well as women may be vendors. As declared by tradition, each stall is an independent business. The board of directors votes on newcomers, although a vacancy is rare.

Margie Denchfield of Bethesda owns the stall her parents took over in the 1960s.

"We were the first city people," she said.

Margie's stall overflows with cakes, pastries, and specialty breads. On neighboring counters, cut flowers and houseplants lend an old-fashioned aura. One showcase houses home-dressed chickens and sausages, another has vintage clothing, and yet another sells used books. In some stalls, ethnic foods simmer in enticingly fragrant spices. In season, there is a wealth of fresh produce to be purchased.

Most of the farms mentioned in Chapter 1 are gone. Charlotte Potter's land is part of the Washington beltway. Her house was moved to safer ground and remains, but her gardens with all the native plants are buried beneath tons of concrete. Charlotte's son, Neal Potter, left a university professorship in economics to take a leadership role in county politics. He was elected to the position of county executive in 1990, and his interest in farmland preservation and orderly growth have made a lasting mark. Another of Potter's achievements was to give strong support for public transportation, resulting in the addition of more and longer trains for commuters into Washington, D.C.

Eleanor Waters's handsome manor house was bulldozed to make way for Germantown expansion, and a large apartment complex was erected on the site. Fronting the building is a nicely landscaped sign with the name "Waters' House."

Jack and Elizabeth Daniel Davis live in a new house on a portion of the Davis family farm. The original farmhouse remains, gracing Thurston Road with its majestic and beautifully maintained facade.

Louise Barnes Duvall is a most ebullient, cheerful, lively octogenarian. Her memory holds stories enough for an entire volume, including a joke she heard at recess in second grade at the Browningsville two-room school. Her old home has been replaced with a new, modern house with no windmill and no Delco battery plant.

Macie King's farm was purchased by the Maryland-National Capital Park and Planning Commission in the late 1960s. Her house, the tenant houses, and the small outbuildings were torn down to make way for the Maryland Soccerplex. But before the bulldozer reached the dairy, Macie King's granddaughter stood in front of the barn and politely said, "No." Barbara McGraw could not bear to see the King Dairy Barn destroyed. She knocked on doors, wrote letters, went to meetings, and spoke her piece for preservation. Grandmother's genes are apparent. Mrs. McGraw is polite but persistent. She is always gracious and smiling, but in her gray-green eyes is steely determination. It's not hard to believe that a Depression-era bank director gave $50,000 to her grandmother's project.

Years went by before McGraw hit pay dirt. At first no one paid much attention to the elegant woman in the perfect clothes who knew just the right thing to say and the right person to say it to. A group formed around her, and local politicians began to take notice. She rented office space in a restored mansion. With her brother, Garner William Duvall; Jack Davis; Carol English; and an efficient board of directors, a plan took shape. The enormous, cement block barn and twin silos would be a living monument to the American dairy farm. Schoolchildren would come to learn about a way of life that was once common but is no more. Antique farm and dairy equipment would find a home and be put on display. The focal point would be the traditional farm work ethic and the production of milk. Each year displays would rotate to lend motivation for visitors to return—one year featuring ice cream, the next year butter, and so on. In a county where both historic and agricultural preservation are highly esteemed, the barn would be an integral part of ongoing county heritage.

McGraw and her board named the project the King Barn Dairy MOOseum.

Little by little, restoration funds appeared. By 2006, over $500,000 had been raised to stabilize the building, paint and repair it, and bring the barn up to county code. Addressing public safety issues is important, and former schoolteacher McGraw understands the need to ensure that no child will be injured.

Grandmother and Grandfather King's dairy barn will be fully restored and open to the public as a living museum and a testament not only to the genius of the dairy farmer, but also to those whose dream it was to preserve this rich history for future generations—Barbara McGraw and the King Barn Dairy MOOseum Board of Directors.

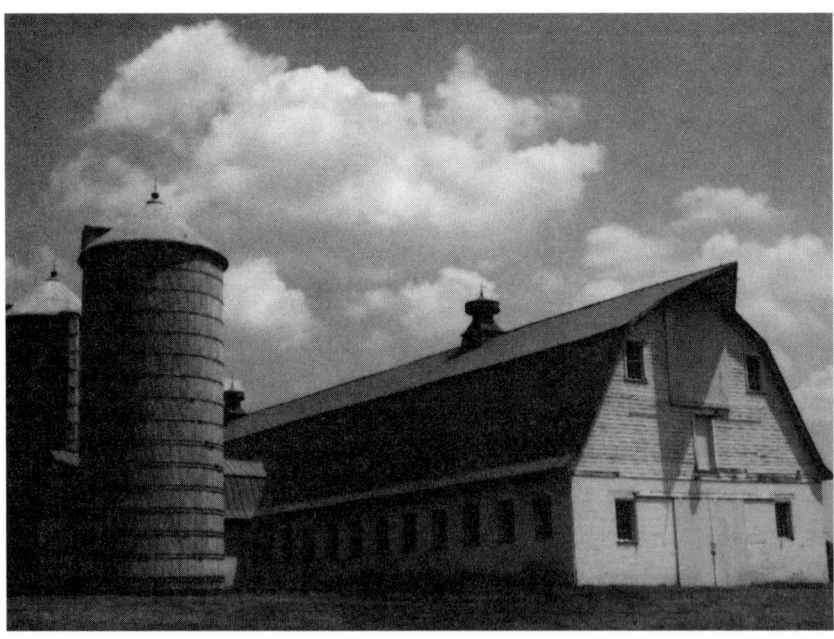

The King Barn Dairy MOOseum. 2007. Photo by the author.

Appendix A
DNA...And the Hum Heard 'Round the World

The day James Watson and Francis Crick discovered the structure of deoxyribonucleic acid is called the Eighth Day of Creation.

On February 28, 1953, Watson and Crick walked into a pub in Cambridge, England, and announced they had found the secret of life. What they had discovered after years of research, both their own and that of other scientists, was the structure of DNA, which is found in the cells of every living thing, whether man, beast, fish, worm, tree, or grass.

Nearly half a century later, Watson and Crick's discovery has been the start of a huge, snowballing industry that is on the cusp of saving lives, improving health, and ending world hunger with more and better food. Disabling health problems such as Alzheimer's, spinal cord injuries, diabetes, Parkinson's, cancer, and many more may be unknown in another generation. DNA is truly a breakthrough in human knowledge, revealing a new era of interpreting our own bodies and our world.

Due to the discovery by Watson and Crick, agriculture is experiencing great change. In the United States, where farmland is shrinking each day, techniques have already been developed to feed more people with less land. Agriculture is undergoing a revolution, and this time women stand with their male associates on a level playing field.

Nearly every day the media announce new discoveries in the realm of genetic engineering. Today's newsmakers are honeybees and prion. Scientific analysis of the DNA of honeybees and their humming may lead to finding causes of human social behavior. Prion is a protein that can cause fatal human disease. It is transmitted by eating beef infected with mad cow disease. Scientists have genetically engineered cattle that are free of prion.

A particularly exciting and controversial result of Watson and Crick's discovery is the ability to clone. In this procedure as it pertains to livestock, a small piece of skin is removed from an animal. The DNA is then analyzed and used to produce an exact duplicate of the skin donor. Viagen, a Texas-based company, has 250 cloned cattle and pigs. At a cost of $15,000 each, they will not be slaughtered for meat, but used for breeding more productive livestock. Hematech, in South Dakota, is cloning animals, too. Hematech's calves are not raised for meat or milk. These cattle will produce substances to be used in making pharmaceuticals for humans.

On other farms, beef cattle are cloned to create meat with the most desirable characteristics; clones of high-producing dairy cows give copious amounts of milk. On December 28, 2006, the FDA announced that there is no risk in eating meat or drinking milk from cloned animals, although voluntary restrictions remain as the government awaited public response to the announcement.

Opponents of genetic engineering distrust the findings of the FDA and fear an unknown consequence from eating cloned or otherwise scientifically altered food. There is concern about the absence of biodiversity. If every cow in the herd is the same, there is no chance for improvement. And the certified organic farmer across the road worries that genetically engineered DNA might be carried by a marauding insect and affect her traditional crop. For example, the gene that carries the natural insecticide BT (Bacillus thuringiensis), it is feared, may be transmitted into a crop certified organic, at a subsequent loss to the organic farmer.

Biotechnology has been around for a long time. The foot-long ears of sweet corn in the market today come from an ancient variety, originally 2 or 3 inches in length. Seed from the 3-inch ears was saved and planted the following spring. At harvest time a 4-inch ear appeared. Its seed was saved and planted, and the process was repeated for centuries, resulting in the current, heftier ear.

Likewise, farm animals show vast improvement. Since they were first domesticated, cattle have been bred for certain traits, resulting in an increased milk supply and better-tasting meat.

Perhaps the earliest account of biotechnology is the biblical Jacob. When told his wages for twenty years of labor were two wives and the "speckled, ringstraked and grisled" cattle, Jacob bred the cows with ringstraked, speckled, and grisled sires, and every calf born was speckled, ringstraked, and grisled. Jacob went off with a very large herd. Biotechnology in the year 1729 B.C.!

Animal biotechnology has been practiced ever since, selectively choosing physical and biological characteristics the farmer wished to see in another generation and arranging for a male carrying these traits to breed the female livestock. For example, biotechnology was responsible for my border collie, a friend's miniature dachshund, and my neighbor's Irish wolfhounds. Every AKC-registered pup has resulted from many generations of intensive breeding.

When artificial insemination was developed for farm animals, frozen semen became easily available by mail order. Controversy arose around a breeding process considered unnatural. Soon the practice was widely adopted, and now a cow may be bred to a sire in North Dakota without ever leaving Maryland. If a farmer needs straighter legs for the progeny of her top-producing cow, she orders semen from a bull whose offspring have straight legs. Herds are improved, and most farmers today practice biotechnology in some form.

Watson and Crick found that the DNA that makes up every cell in every living thing has a marker that identifies a characteristic that can be isolated and transferred. Various names apply to different methods of achieving a desired result. To produce an identical twin or clone of a prized cow, a small piece of skin is removed and the cells identified and isolated. The nucleus with the desired DNA is extracted and implanted in a fertilized oocyte, or cell that has had its nucleus removed. The resulting embryo is placed in any healthy cow. The surrogate will bear a calf with none of her own DNA, but with DNA that is identical to the prized animal—a clone.

A transgenic animal results from a different technology. For instance, consider dairy goats. I happen to love dairy goats. For thirty years I milked my Saanens twice a day and made cheese, ice cream, and rice pudding. My goats were a part of the family. Each one had an individual personality and a distinct sense of humor. My memory bank includes Ingrid, who sailed gracefully over all civilized barriers. Hence my gates and fences grew to absurd heights. Then there were Sally and Chantilly and Lacey and beautiful, borrowed Winnie—each with individual personalities but all with delicious Saanen milk.

However, there's more goats can do besides provide milk for cheese!

In Massachusetts lives a transgenic goat named Sweetheart who produces lysozyme, a protein found in human breast milk, which prevents disease by increasing antimicrobial activity.

Sweetheart's caregivers are not making ice cream, but ATryn, "the first drug produced by a transgenic animal to be approved by the European Medicines Agency." ATryn is useful in treating acquired antithrombin deficiency, which can result from burns, disseminated intravascular coagulation, sepsis, cardiopulmonary bypass surgery, acute liver failure, and other conditions.

Sweetheart lives on a farm run by GTC Biotherapeutics, where she enjoys the position only great success can bring. If there is a Nobel prize for goats, Sweetheart will surely qualify.

A transgenic goat (or sheep, cow, pig, rabbit, or chicken) results from pronuclear microinjection, whereby a new gene, or transgene, is implanted into a freshly fertilized oocyte using a very fine glass needle. From then on, the same technique as the one used in cloning is followed. The embryo is placed in a petri dish and observed for cell development before being transferred—in the case of goats—to a recipient doe to grow into a full-term kid.

Undoubtedly Sweetheart's birth mother was cared for in good goatherd practice, and five months later Sweetheart was born. She was raised and bred in normal fashion. When she freshened, her milk contained the protein needed to manufacture ATryn.

Exciting news gushes from science laboratories working with human health. By using compounds in milk from genetically modified goats, cows, and pigs, many human ills may soon be forgotten. At the University of Nebraska, researchers are growing genetically engineered pigs whose milk contains the proteins that are deficient in people with hemophilia. For use in human health, the transgene is removed from one of three sources: excess fertilized eggs (ova) that result from infertility treatment; an amniotic stem cell that is produced in cells shed naturally in the amniotic fluid of a pregnant woman; and adult stem cells from adult bone marrow.

The cells develop into embryos, which may be shipped to other farms and other countries. DNA marker-assisted selection is apparently the final step to producing genetically superior animals. The embryo in the petri dish may be mailed to Tanzania or Timbuktu or Barnesville, Maryland, where superior stock improves herds exponentially.

A host of geneticists are at work throughout the world conducting experiments using human stem cells to produce farm animals that produce more milk, and more and better-tasting meat, to feed a hungry world, and to produce drugs that will cure the diseases that cause pain and take the lives of human beings. Geneticists are also studying the genes of grain-bearing plants to grow grain that is more nutritious and more plentiful. At the same time, farmers may achieve their economic potential by selling products to pharmaceutical firms.

The Johns Hopkins University Belward Research Campus is located on the former Banks farm. If only Elizabeth Banks could see what is happening on her farm! Liz cried when her cows left Belward, but ironically, the current owners may make a difference far beyond her imaginings. Although it is unlikely that Liz Banks paid much attention to the announcement by Watson and Crick, there is reason to hope that society will be healthier and live happier, longer lives with products now being developed on Liz's fertile pastures.

The Johns Hopkins University Belward Research Campus is the home of Human Genome Sciences (HGS), a pharmaceutical firm that offers hope to millions. As a pioneer in genomics research, HGS is developing products that may make "fundamental improvements in human health worldwide," said former Maryland Governor Parris Glendening at the opening of the firm in 1999. Recognized as the first facility in the world designed to create medicine from the study of genes, Belward is manufacturing genomics-derived drugs. In October 2006, HGS at Belward introduced Albuferon, the first such product to be accepted by the FDA and offered for sale in the United States. Albuferon may prove useful in cancer research and in controlling aging. The drug is currently undergoing clinical trials with hepatitis C patients.

Paralleling the Human Genome Project came the study of genetics in farm animals. This was exciting news to farmers barely able to pay their feed bill. Using the new technology, perhaps they can make money on their farms without having to sell the land for development. Production costs are high, but today, embryos from a prime dairy cow may sell for as much as $30,000.

Universities with animal science departments responded to Watson and Crick with classes of bright young scientists eager to harness this new technology. Change occurred. In the 1950s, women were not allowed in the faculty coffee room of most university science departments. Today women science students equal men in numbers and in skills.

In the early 2000s possibilities seemed endless, and the government was quick to realize not only the benefits to our nation's health resulting from the genome project, but also the importance of keeping talented American scientists here at home.

Hopes were battered in August 2001, when President George W. Bush withdrew support of stem cell research. He announced that federal money would be used only for research with embryos that had been destroyed by the time of his speech. Many research scientists lost their funding as a result.

The Netherlands and European countries have established laboratories to make full use of DNA studies and are luring bright doctoral candidates with the offer of high salaries and opportunities for new discoveries.

On July 19, 2006, President Bush renewed his decision with the first veto of his administration. The Senate had passed a bill providing more federal funding for research, but the House of Representatives fell short of the two-thirds majority needed to override the veto.

Individual states provide funding, however, and the excitement swells.

On November 20, 2007, scientists James Thomson of the University of Wisconsin and Shinya Yamanaka of Kyoto University announced the successful results of their research. Adult skin cells can be developed into replacement parts for the donor, including nerve, heart, muscle, and other tissues. President Bush subsequently said he will support this research.

The science of genetics has exploded, and both women and men are entering the profession. Nature and other scientific magazines abound with ads for geneticists, particularly in other countries. Some state a preference for female applicants.

Farm women still need two aprons. This time, one is a lab coat.

Appendix B
Government and Agriculture

During the Great Depression of the 1930s, the federal government heard the cries for help and held out a hand to farmers. The extension service provided education concerning growing better crops, but farm marketing needed help.

The Agricultural Marketing Act of 1929 attempted to set a floor beneath the price level for major export crops. President Herbert Hoover signed the bill, which guaranteed that the farmer would get no less than a predetermined price for crops. This policy might have worked in a good economy, but worldwide depression meant that prospective buyers couldn't afford the fixed prices.

In the early 1930s, dust from the drought in mid-America darkened skies from Texas to the Chesapeake Bay. Responding to the disaster, Congress and President Hoover signed the Soil Conservation and Domestic Allotment Act, creating the Soil Conservation Service (SCS). When Franklin Delano Roosevelt took office in 1933, his Agricultural Adjustment Act (AAA) continued the alphabet soup of government programs designed to rid the nation of depression and increase living standards for farmers. The bill authorized production adjustment programs, and created a license that was required in order to promote orderly marketing of fruits and vegetables. Some farmers saw the license as more red tape that would slow production. The AAA also offered loans for farmers. A wide range of initiatives was introduced, including the publication of Consumer's Guide. Charlotte Potter's copy of the Guide contains an early feature story on the Montgomery County Farm Women's Market.

In 1936 the Supreme Court declared major portions of the first Agricultural Act unconstitutional, and the legislation was replaced with the Agricultural Adjustment Act of 1938. The new regulations authorized mandatory supply controls through acreage allotments and marketing quotas.

World War II provided innovations useful to farmers. Flies and rats were exterminated with DDT and a substance referred to as 1080, "a powerful new poison developed during the war. Experts will be brought in to handle this poison which is extremely dangerous if not properly used."

Using Jim and Macie King's farm for demonstration purposes, county agricultural agent O. W. Anderson contoured fields with a series of furrows climbing to the highest point on the property. From there, water would be pumped and allowed to run through the furrows to all sections of the farm, ending the lingering fear of another drought.

Agent Anderson needed an assistant, so Bob Raver, an agricultural scientist, joined the Montgomery County Extension Service in 1957.

During the 1960s land speculators bought open space and farmland like kids cleaning up the chocolate-icing bowl. "I've never seen anything take place so fast as [the sale of farmland in] Montgomery County in the '60s," Raver said.

Taxes were negligible on land assessed for agriculture, enabling investors to buy the land for future development and ignore the ground until it was sold. But Maryland's state legislature passed a law that forced owners to farm the land in order to receive the tax advantage. Hiring custom-farming contractors became common practice. To prepare for spring planting, the contractor worked the soil vigorously, tilling and disking deep into the ground before planting. When spring rain came, "tons of topsoil were lost to erosion," Raver said.

Raver was concerned about the loss of topsoil and also about silt buildup in the waterways. He studied the matter and devised a method to farm that would not disturb the soil, called no-till, and announced a demonstration for local farmers. Raver sprayed the land with herbicide, then planted corn directly into the soil. No weeds competed with the crop, and no loosened soil silted the streams.

The resulting conservation of water, soil, and manpower was dramatic. No-till "took off like wild fire," said Raver. Within four years 85 percent of Montgomery farmers grew their corn using the no-till method.

In the years before ethanol, farmers needed a market for their corn and looked to the federal government for help. Although frequently revised and amended, the Agricultural Adjustment Act is still in effect. In 1980 the federal administration established a grain embargo with the Soviet Union, which caused consternation in countries importing food from the United States. Questioning the reliability of the United States as a food source and criticizing the price support program as an unfair subsidy, some nations looked around for other places to buy grain.

In 1985 the Omnibus Food Security Act spelled the end of family farming in many communities and may have been a major contributor to the establishment of agribusiness. Proposed as a way to make the United States competitive in the world market, the federal government set up a whole-herd buyout program in exchange for a commitment from the farmer to stay out of the dairy business for at least five years.

Observing the surge of farmland loss to development coupled with high prices offered by investors, some farmers found the offer of cash for their herds irresistible.

In Minnesota, agribusinesses bought up dairy farms, and today thousands of acres are worked by one owner. Corn for ethanol has replaced cows as the preferred moneymaker. According to a native of Sturgeon Lakes, Minnesota, family farms and entire villages, schools, and churches were abandoned as one agricultural corporation took over all the farmland in the area.

By 2006, the federal government was paying $15 billion in annual farm subsidies. The largest farms got the most money, enabling them to buy more and more land and crushing small family farms. The law authorizing these payments expired in 2007, and a new farm policy is under study.

Sources

Books

Buchanan, Rita. *A Weaver's Garden*. Loveland, CO: Interweave Press, 1987.

Coleman, Margaret M., and Anne Lewis. *Montgomery County: A Pictorial History*. Virginia Beach, VA: Donning Co., 1983.

Encyclopedia Britannica, 1959.

Farquhar, William Henry. *Annals of Sandy Spring*, vol. 1. Baltimore: Cushings & Bailey, 1884. Reprint, Cottonport, LA: Polyanthos Inc., 1971.

Goodall, Jane, with Gary McAvoy and Gail Hudson. *Harvest for Hope: A Guide to Mindful Eating.* New York: Time Warner, 2005.

Jones, LuAnn. *Mama Learned Us to Work: Farm Women in the New South.* Chapel Hill: University of North Carolina Press, 2002.

Judson, Horace Freeland. *The Eighth Day of Creation: Makers of the Revolution in Biology*. Plainview, NY: Cold Spring Harbor Laboratory Press, 1996.

MacMaster, Richard K., and Ray Eldon Hiebert. *A Grateful Remembrance: The Story of Montgomery County, Maryland.* Rockville, MD: Montgomery County Historical Society, 1976.

Maryland-National Capital Park and Planning Commission. *Preservation of Agriculture & Rural Open Space*. Silver Spring, MD, 1980.

Miller, Kristic. *Isabella Greenway: An Enterprising Woman*. Tucson: University of Arizona Press, 2004.

Offutt, William M. *Bethesda: A Social History*. Bethesda, MD: The Innovation Game, 1996.

Panno, Joseph. *Animal Cloning: The Science of Nuclear Transfer*. New York: Facts On File, 2005.

Parkinson, Mary Jane. *And Ride Away Singing: The Breeding Philosophy of Bazy Tankersley and the History of Al-Marah Arabians*. Tucson, AZ: Al-Marah Arabians, Arabian Horse Owners Foundation, 1998.

Shortall, Sally. *Women and Farming: Property and Power.* Houndmills, Basingstoke, Hampshire: Macmillan Press; New York: St Martin's Press, 1999.

Smith, Richard Norton. *The Colonel: The Life and Legend of Robert R. McCormick, 1880-1955*. New York: Houghton Mifflin, 1997.

U.S. Government Printing Office. *A Brief History of the Committee on Agriculture and Forestry, United States Senate and Landmark Agricultural Legislation, 1825-1986*. Senate Print 99-213. Washington, D.C.: 1986.

Periodicals, Journals, and Archives

Bethesda Journal, Bethesda, MD. 1940-1950. Archives of the Montgomery County Historical Society.

Brooke, Mary Briggs. *Diaries of Mary Briggs Brooke*, ca. 1868-1890.

Carrell, Jennifer Lee. "They Drink the Wind." *Smithsonian*, vol. 20, no. 6, p. 49.

Eenennaam, Alison L "What is the future of animal biotechnology?" *California Agriculture*, vol. 60, no. 3, July-Sept, 2006, p. 136.

Funk, W. C. *North Dakota Farmer*. Read at USDA, Beltsville, MD, June 1916. Out of print.

Grantsmanship Center, Los Angeles, CA 90017. http://www.tgci.com.

Hedgpeth, Donna. *The Washington Post*, January 24, 2005, p. B3.

High, Stanley. "Country Kitchen Goes to Town." *Reader's Digest*, vol. 36, no. 213 (January 1940), pp. 95-96.

Holstein-Friesian World, May 10, 1970, April 10 1976. Archives of Dairy Business, 6437 Collamer Road, East Syracuse, NY 13057-1031.

Honeybee Genome Sequencing Consortium. *Nature*, October 26, 2006, pp. 931-949.

Human Genome Sciences Press Release, Rockville, MD, April 14, 1999.

King, Macie Schaeffer. Scrapbooks circa 1922-1960. Courtesy of Garner W. Duvall Jr., Rockville, MD.

Ledford, Heidi. "The Farmyard Drug Store." *Nature*, September 7, 2006, pp. 16-17.

Montgomery County Extension Service Archives, Agricultural Farm Park, Derwood, MD.

Montgomery County Maryland Land Records and Incorporation Papers, Rockville, MD. Elias D. King to James D. King; 324 160, December 9, 1922.

CKW 1 p. 92. Incorporation papers, Montgomery County Women's Cooperative Farm Market.

Nelson, Jerry. "Bioherds=big bucks: transchromosomic cows may help our health and your wallet." *Dairy Today*. August 1, 2003. Farm Journal Media. Vol. 19, issue 7, p. S14.

Paarlberg, Don. "Effects of New Deal Farm Programs on the Agricultural Agenda a Half Century Later and Prospect for the Future." *American Journal of Agricultural Economics*, vol. 65, no. 5 (December 1983), p.

1163.

Rehmeyer, Julie J. "Milk Therapy, Breast-milk compounds could be a tonic for adult ills." *Science News*, vol. 170, no. 24 (December 9, 2006).

Richardson, Anna. "Gourmets Gather Here." *Woman's Home Companion*, October 1937, p. 27.

Time magazine. "A Castle for the Princess." April 15, 1949.

Time magazine. "The Most Important People of the Century." March 29, 1999, p. 1.

U.S. Agricultural Adjustment Administration. *Consumers' Guide*, vol. 1, no. 11 (1934).

USDA Archives, National Agriculture Library, Special Collections Room 300, 10301 Baltimore Ave., Beltsville, MD 20705.

USDA Farmer Cooperative Service. News for Farmer Cooperatives, August 1957.

Washburn, Ashley. "A Swine Sensation: Protein from the Milk of Transgenic Pigs Could Be a Revolutionary Treatment for Hemophilia." *Engineering Nebraska Magazine*, Spring 2006. http://www.engineering.unl.edu/ENonline/Spring 06/18.shtml.

Washington Post. "Frankenfood? Not Quite." Editorial, February 10, 2007.

Weiss, Rick. "FDA Is Set to Approve Milk, Meat from Clones," *Washington Post*, October 17, 2006.

———. "FDA Says Clones Are Safe to Eat," *Washington Post*, December 29, 2006.

———. "Scientists Announce Mad Cow Breakthrough," *Washington Post*, January 1, 2007.

———. "Scientists See Potential in Amniotic Stem Cells," *Washington Post*, January 8, 2007.

———"On Stem Cell Legislation, a Reprise with Twists," *Washington Post*, January 11, 2007.

Wiser, Vivian D. "Women in Agriculture." Paper presented at USDA conference, January 1976; archival files of USDA.

Women's Mutual Improvement Association, vols. 1-10, 1855-1916.

Conversations and Interviews

Cunningham, Donovan. January 1, 2007.

Cunningham, Eleanor. Asbury Retirement Village, Gaithersburg, MD. August 23, 2006.

Davis, Elizabeth and Jack. Thurston Rd., Dickerson, MD. 2005 and 2006.

Denchfield, Margie. Bethesda, MD. January 11, 2006.

Duvall, Garner W. Rockville, MD. January 18, 2006.

Duvall, Louise Barnes. Damascus, MD. December 11, 2006.

Foster, Delbert. Extension Service (ret.). Gaithersburg, MD. March 20, 2006.

Geary, Diane Savage. Agricultural Farm Park, Derwood, MD. January 25, 2006.

Hardisty, Debbie. Berryville, VA. April 10, 2006.

Johnson, Peggy Dayhoff. Dickerson, MD. January 9, 2006.

Jones, Cecile King. Cedar Grove, MD. January 10, 2006.

Kellerman, Frances Wilmot. Gaithersburg, MD. January 17, 2006.

Kingsbury, Peggy Horine. Kingsbury Orchard, Peach Tree Rd., Dickerson, MD. 2005-2007.

Lechlider, Caroline and George. Laytonsville, MD. September 11, 2006.

McGraw, Barbara Duvall. Bethesda, MD. January 6, 2006, and other dates.

Moxley, Susan, DVM. Frederick, MD. May 3, 2006.

Nicholson, D. Boyds, MD. 2005-2006.

Phillips, Jean King. Schaeffer Rd., Germantown, MD. Winter 2005.

Potter, Marian and Neal. Brookeville Rd., Chevy Chase, MD. 2005 and 2006.

Raver, Robert. Mt. Ephraim Rd., Dickerson, MD. January 31, 2006.

Schwartzbeck, Nona. Union Bridge, MD. February 28, 2006.

Sopko, George, M.D. Boyds, Maryland. 2007.

Stamberg, Elena. Barnesville Rd., Barnesville, MD. January 19, 2006.

Stiles, Barbara Ann Riggs. Charles Town, WV. 2005-2007.

Stiles, Janet. Boonsboro, MD. November 15, 2006.

Stup, Olive Gladhill. Damascus, MD. January 2, 2007.

Tregoning, Douglas. Extension Service Agent. Microsoft PowerPoint CD, "Women in Agriculture," and conversations, 2006-2007.

Whirley, Laura Ann. Derwood, MD. March 2006.

Afterword

There were no women in his life, but a book featuring Maryland dairy farms in the 1960s to '70s would be incomplete without a nod to Paclamar Astronaut. "He's rewriting the record books," announced *Holstein-Friesian World*, April 10, 1976.

Purchased by a consortium of Montgomery County farmers and one West Virginian on June 11, 1964, Astronaut became the youngest bull to earn his Gold Medal award in artificial insemination, achieving the honor in 1969. By 1976 he had 39,670 registered sons and daughters! They produced a lot of milk. Astronaut's 23,101 tested daughters gave an annual average of 8,000 quarts of milk. Astronaut's delighted owners included several Kings—Billy, Jack, Merhle, and James W.—and Charles Savage, W.L. McDanolds, Richard Tarbutton Jr., Leroy Savage, and W. J. Greer. SOURCE: *Holstein-Friesian World*, May 10, 1970, and April 10, 1976; archives of Dairy Business, 6437 Collamer Road, East Syracuse, NY 13057-1031.

INDEX

Note: Page numbers in *italics* indicate photographs.

Adams, Bruce, 103
Agribusiness
  dairy farms sold to, 118
  farm subsidies for, 118
Agricultural Adjustment Act, 115-16, 117
Agricultural Marketing Act of 1929, 115
Al-Marah (farm), 85-89
American Board of Agriculture, 7
American Heart Association benefit pageant, 87-88
Anderson, Otis W., 22, 52, 116
Annals of Sandy Spring, 4
Arabian Museum, 86
Arnold, Leon, 55, 60

Banks, Elizabeth "Liz" (Beall), 89, 90
Barnes, Dorothy, 31, 33
Barnes, Harriet, 31, 33, 64
Barnes, Herbert, 31, 32-33, 63
Barnes, James Oliver, 31, 33
Barnes, Mazie, 31, 33
Barnes, Rosa, 31, 33, 42, 49
Barnes, Vivian, 31, 33
Barnesville (Maryland), 85, 86, 87
Belward (farm), 89-91
Belward Research Campus, 112-13
Bentley, Caleb, 5, 6
Bentley, Sallie Chandlee, 12
Bentley, Sarah (Brooke), 5
Bethesda (Maryland)
  government agency campuses in, 77, 82
  See also Edgemoor
Biotechnology
  artificial insemination, 109-10
  in Biblical times, 109
  genetic engineering, 107-09, 112-13
  stem cell research, 112, 113-14
  transgenic animals, 96, 110-12
Bird, Jacob, 39
Bomberger, Frank, 61

Boonesboro (Maryland), 93
Briggs, Eliza, 12
Briggs, Hannah (Brooke), 5, 7
Briggs, Isaac, 5, 7
Brink (farm), 23, 73
Brooke, Deborah (Snowden), 5
Brooke, James, 5
Brooke, Mary (Briggs), 11-12
Brooke, Roger IV, 5
Brookeville (Maryland), 4, 5
Browningsville (Maryland), 31
  schoolhouse, 33, 104
Bush, George W., 113, 114
Butter-making, 6-7

Cattle farming, 90
Cedar Grove (Maryland), 23
Central Bank of Monrovia, 42
Dime Savers, 33, 42
Charlotte's Filled Cookies (recipe), 70
Chicago Tribune, 83
Civil War, in Montgomery County, 11
Clagett, Mrs. Chester, 56
Cloth
  dyes for, 8-9
  feed sacks used for, 2, 39, *94*
  silk, 8
  wool, spinning of, 18
Conoy (Barnesville neighborhood), 87
Consumers' Guide (newsletter), 116
Contoured plowing, 116
Cooperative Bank (Baltimore), 61
Cooperative Extension Service. *See* Homemakers Clubs; Montgomery County Extension Service; University of Maryland Extension Service
Cooperative Marketing Association, 19
Corwin, Blanche I., 19, *21*, *22*, 25, 41, 42, 43, 74
  farm women's market and, 45-47, 51-52

fired from job, 52
home demonstration agent work, 34-36
teaching about marketing, 46
Cows. *See* Cattle farming; Dairy farming
Crick, Francis, 107

Dairy farming, 6-7, 37, *66*, 73, 99
  bottling procedures, 27
  breeding, 95
  butter, 6-7
  genetics and, 96, 99
  health & sanitation, 27, 28, 99
  milk bottles, 27, 37, *38*
  milk delivery, 23-24, 30, 74-75, 95
  milk ordinance, 22
  modern practices, 93, 95-96
Dairy farms, agribusiness purchases of, 118
Damascus Homemakers Club, 42
Damascus (Maryland), 31, 33
Daniel, Elizabeth "Boo". *See* Davis, Elizabeth "Boo" (Daniel)
Daniel, Elsie (White), 30, 57, 70
Daniel, Mansfield, 30
Daniel, William, 30, 31
Daniel, William Jr. "Billy," 30
Darby, Mrs. John, 67
Davis, Elizabeth "Boo" (Daniel), 30, 47, 60, 104
Davis, Harriet, 28
Davis, Jack, 28, 60, 104, 105
Davis, John, 26, 28, 47, 65
Davis, John Wallace, 27, 28
Davis, Leonard Isaac, 28
Davis, Margaret (formerly Magruder), 26-27, 28, 37, 47, 57, 60, 65
Davis, Peggy, 28
DDT, 116
Delco battery plant, 63-64, 104
Depression era. *See* Great Depression
District of Columbia
  Georgetown, 6

  See also Washington, D.C.
Drought, in midwestern states, 115
Duvall, Garner William, 105
Duvall, Louise (Barnes), 31-32, 33, 49, 63, 104

Edgemoor (Bethesda neighborhood), 46, 51
  former Piggly Wiggly store, 51, 52
  See also Bethesda
Electricity, 63, 64-65
English, Carol, 105
Enterprise Farmers Club, 7
Extension Service. *See* Montgomery County Extension Service; University of Maryland Extension Service
Extension Service agents, 15-16, 18, 101, 102-03
  See also names of specific agents
Extension Service Homemakers Clubs. *See* Homemakers Clubs

Farley, Walter, 87
Farm labor, 24
  contractors for land investors, 117
  farmerettes, 18
  post World War II, 16, 75, 77, 82
  prisoners-of-war, 75, 77
Farm wives, 3-4
  clothing, from feed sacks, *2*
  daily life, 1, *2*, 10-11, 16, 49, 56-57
  See also Women's status
Farm Women's Cooperative Market. *See* Montgomery County Farm Women's Cooperative Market, Inc.
Farmerettes, 18
Farmer's Non-Partisan Taxpayers League, 37
Farmhouses
  electrical power, 63, 64-65
  modern conveniences in, 47, 65
  running water, 27, 63-64
Farming
  added-value, 99

rising cost of, 99-100
Farmland
    eminent domain and, 82
    prices, 82
    woodland loss, effect on ecology, 90
Farquhar, William Henry, 4-5
Federal Silk Company, 9
4-H Clubs, 77-79

Gaithersburg (Maryland), 79
Geary, Diane Savage, 98-100
Genetic engineering. *See*
    Biotechnology
Germantown (Maryland), 23, 33, 104
    Liberty Mill, 24
Gilpin, Rachel, 12
Gladhill, Lillian, 33
Gladhill, Molly, *69*, 102
Glendening, Parris, 113
Goats, 110-11
Godfrey, Arthur, 80
Graceful Legend Llamas (farm), 97-98
Greary, Diane Savage, 3
Great Depression, 4, 37-42, 43, 46, 49, 62, 73, 77
    abandoned children, 39-41
    farmers' protests, 37
Greenway, Isabella, 88

"Hang Up the Baby's Stocking" (poem), 25-26
Hanna, Mark, 83
Hargett, Mary, 67
Hargett, Nellie, 65, 68
Henderson, Ida Mae, 34, 47
Higgins, H. M., 25
Hilton, Sarah Elizabeth Brown, 2
Home Demonstration Agents, 19, 25, 34-36, 101
Homemakers Clubs, 25, 30, 34, 41, 42, 43
Hoover, Herbert, 115
Horse farms, 85-89
Horses, 24, 30, 85

Housing developments. *See* Urban sprawl
Human Genome Sciences (HGS), 112

Ice Box Rolls (recipe), 70-71
Indian Pudding (recipe), 15
Indigo, 8-9
Indoor plumbing, 27
Indraff (horse), *84*, 85
Influenza epidemic, 18
Inverness (farm), 30
Irvington Farm, 75

Jacobs, Betty Jean (King), 75
Jacobs, James Wriley, 3
Jefferson County Farm Tour, 98
Jefferson, Thomas, 5
The Johns Hopkins University, 90-91
    Belward Research Campus, 112-13
Johnson, Peggy, 81
Jones, Cecile (King), 77
Jones, Elizabeth S., 19, *20*
Jones, Mary Harris "Mother," 49-50

Kellerman, Frances (Wilmot), 58, 60, 79, 80, 81
King Barn Dairy MOOseum, 105-06, *106*
King, Betty (Birgfield), 73, 74
King, Elias Dorsey, 23, 77
King, Elizabeth "Betty" (Fulkes), 74, 75-76
King, Helen Gertrude, 24
King, James D., 23, 27, 116
King, James S., 24, 73
King, Lawson, 74, 75, 76, 79, 82
King, Macie Irene, 24
King, Macie (Schaeffer), 23, 25, 27, 41, 45, 47, 57, 61, 79, *88*, 104, 116
    as president of the Farm Women's Cooperative Market, 53, 55
    slaughtering license, *59*
King, Merhle, 73, 74
King, Pearl, 45, 57

Kingsbury, Peggy, 39

Lea, Deborah, 10
Lechlider, Caroline, 3
Lee, E. Brooke, 51, 52
Liberty Mill, 24
Linthicum, Arthur, 81-82
Linthicum, Earl, 32
Linthicum, Ethel, 81
Llamas, 97-98
Locke, John, 3-4
Lone Ranger, 87
Longworth, Alice Roosevelt, 85

Madison, Dolley, 6
Madison, James, 5
Magruder, Margaret, 26
   See also Davis, Margaret
Magruder, Thomas, 26
Main, Ada, 74
Maryland, See also Montgomery County; names of specific towns and places
Maryland Soccerplex, 105
McCormick, Medill, 83
McCormick, Robert R., 83
McCormick, Ruth (Hanna), 83
McGraw, Barbara, 105
Milk
   from genetically modified animals, 111-12
   in schools, 96
Miller, Kristie, 85
Miller, Mark, 85
Mince Meat (recipe), 29
Minnesota, agribusiness in, 118
Montgomery County, 4
   Civil War actions in, 11
   government agencies in, 77, 82
   IBM campus, 82
   land speculation, 116-17
   land-use planning, 4, 82
   loss of farmers in, 99-100
   milk ordinance, 22

   urban development, 4, 81, 82, 90, 104-05
   See also Maryland; *names of specific towns and places*
Montgomery County Agricultural Society, 7
Montgomery County Extension Service, 77-79, 98
Montgomery County Farm Bureau, 24
Montgomery County Farm Women's Cooperative Market, Inc., 44, 51, 54, 62-63, 103
   in 21st century, 103-04
   building, 55, 60, 61
   cleanliness emphasis, 46, 53, 56
   creation of, 43-46
   customers, 60, 61-62
   dress code, 47, 55, 103
   effects on farm lifestyle, 49, 56-58
   goods sold at, 57-58, 61, 62-63, 65, 67-68, 103-04
   incorporation, 56
   membership rules, 56
   objections to, 51
   opening day, 46-47
   origins, 43-46
   price competition, 56
   refrigeration, 47, 56
   success of, 47-48, 51, 53, 62, 65, 67-68, 73
Montgomery County Home Demonstration Agent, 19, *21*
Montgomery County Silk Company, 9
Montgomery Farmers Club, 7
Moore, Clayton, 87
Moore, Estelle, 14
Moore, Mary (Brooke), 5
Moore, Thomas, 5, 6-7, 14
Mother Jones, 49-50
Moxley, Susan, 89, 90

National Institutes of Health, 77

Oden, Donna, 80-81

Omnibus Food Security Act, 117-18
Orchards, 18
Orphan Trains, 39-40

Pesticides, DDT, 116
Philadelphia Agricultural Society, 6
Pierce, Sally, 10
Pig Club, 19
Pigs, 19, *20*
　bladders as toys, 12
Pinckney, Eliza Lucas, 9-10
Poolesville Fair, Women's Department, 19
Poolesville (Maryland), 30
Potter, Alden, 29-30
Potter, Charlotte (Waugh), 29, 30, 41, *50*, 52, 57, 68, 104
Potter, Lloyd, 29, 30
Potter, Neal, 29, 30, 57, 103, 104
Poultry farming, 30-31
Prisoners-of-war, as farm laborers, 75, 77

Quaker society, farm women in, 5
Quakers, 19, 101
　Cooperative Marketing Association, 19
　tradition of writing, 4
Questers club, 57-58

Rabbitt, Herman, 89-90
Raver, Bob, 116, 117
Ray, Nannie, 67
Reader's Digest (magazine), 67
Recipes
　Charlotte's Filled Cookies (recipe), 70
　Ice Box Rolls, 70-71
　Indian Pudding, 15
　Mince Meat, 29
　Red Flannel Hash, 14-15
　Valley Forge Bullets, 68
　Red Flannel Hash (recipe), 14-15
Refrigerator, portable, invention of, 6

Richardson, Anne, 62-63
Riggs, Barbara Ann, *78*, 79
Riggs, Joyce, *78*, 79
Roberts, Mary L., 9, 10, 11
Rockville (Maryland), 18, 40, 41, 51, 52, 90
Rockville (MD) Sentinel, 51, 52
Roosevelt, Eleanor, 62
Roosevelt, Franklin Delano, 115
Rural Electrification Administration, 64
Ruth McCormick Tankersley Charitable Trust, 88

Sandy Spring Farmers Club, 7
Sandy Spring Farmers' Convention, 37
Sandy Spring Farmers Society, 7
Sandy Spring (Maryland), 4, 6*, 12, 14, 101
Savage Farm, 99
Schaeffer, Allen D., 24
Sharecroppers, 34
Sheep, 18
Shenandoah Jerseys (farm), 93-96
Sherwood (farm), 9
Shorb, Mary, 65
Silkworms, 7-8
Silver (horse), 87
Silver Leaf Flour, 24
Simms, Albert, 83
Slaughterer's license, *59*
Slaughtering, 81
　of hogs, 11-12
　outlawed in Montgomery County, 81
Smith-Lever Act of 1914, 15-16
Snarr, Wardney C., 18-19, 22, 35
Social Service League, 39
Society of Friends. *See* Quakers
Soil conservation
　contoured plowing, 116
　no-till farming, 117
Soil Conservation and Domestic Allotment Act, 115
Stabler, James P., 7, 8, 9

Stabler, Sarah Briggs, 7, 8, 9
Stamberg, Elena, 97-98
Steiner, Merle, 91
Stiles, Janet, *92*, 93-96
Stiles, Mrs. Grace, *94*
Stiles, Tracy, 96
Sturm, Anne (Tankersley), 85
Sturm, Joanna, 85
Sugarloaf Mountain (Maryland), 26
"The Summer Boarder" (poem), 13-14
Summer boarders, 12
Summit Hall (farm), 79, 80-81
Sweetheart (goat), 110-11
Symons, T. B., 35, 52

Tabb, Jane, 98
Tankersley, Garvin, 85, 86, 88
Tankersley, Garvin Jr., 85
Tankersley, Ruth "Bazy" (McCormick), 83-89
Tankersley, Tiffany, 87, 88
Telephone demonstration, *21*
Thomas, Lydia, 10
Thompson, James, 114
Thompson's Dairy, 74
Time (magazine), 83
Tobacco farming, 31-33, 37
Toys, pigs' bladders as, 12
Travilah (Maryland), 34
Tregoning, Douglas, 102-03
Triadelphia (Maryland), 7
Trigger (horse), 87
Turf farming, 79, 80
Turner, Edythe, 52, 53

University of Maryland, Rural Women's Retreats, *22*, 35, 101
University of Maryland Extension Service, 15, 22, 101
University of Nebraska, 111
Urban development, 4, 81, 82, 90, 104-05
U.S. Department of Agriculture
Cooperative Extension Service, 15
Soil Conservation Service (SCS), 115
See also Homemakers Clubs; Montgomery County Extension Service; University of Maryland Extension Service
U.S. Supreme Court, 116

Valley Forge Bullets (recipe), 68
Van Hosen, Fred J., 16-17
Vegetables
gardens for, *17*
potatoes, 33
preserving of, 16

Wade, Sarah, 39
Washington, D.C., 4
Washington County, milk in schools, 96
Washington Post, 91
Washington Times-Herald, 83, 85
Waters, Eleanor (Cissel), 33, 41, 62, 63, 65, 104
as president of the Farm Women's Cooperative Market, 60-61
Watson, James, 107, 110
West, Mrs. Edward, 67
Wheat, 18, 24
grain embargo with Soviet Union, 117
Wheaton (Maryland), 82
Wilmot, Bill, 80
Wilmot, Frances. See Kellerman, Frances (Wilmot)
Wilmot, Frank, 58, 79
Wilmot, Joe, 58, 79
Wilson, Woodrow, 15
Wiser, Vivian D., 1
Woman's Home Companion (magazine), 62, 67
Women's Mutual Improvement Association, 9-10, 14
Women's Mutual Improvement Association journals, 10-12
Women's status, 3-4, 50-51, 60
changes in 1960s, 77-79

commercial property purchasing by, 60
Depression effect on, 49-51, 77
farmerettes, 18
suffrage, 14
in university science departments, 113
See also Farm wives
World Dairy Exposition, 93
World War I, support for troops, 16
World War II, 75, 116

Yamanaka, Shinya, 114

About the Author

All her life, Margaret Coleman wanted to be a farmer. In 1980, she and her husband Jim bought 30 acres of farmland and a broken-down cabin in Montgomery County, Maryland. A farmer at last, Margaret got to work herding sheep and planting a large garden. She takes her herbs and flowers to market and sells dyed wool and "shepherd's whimsies" out of her restored cabin which she has turned into a successful farm bed-and-breakfast called Pleasant Springs Farm. Her four children and eight grandchildren often visit her rural home for a chance to get away from the hustle and bustle of city life. Not one to be outshone by her multi-faceted ancestors, Margaret is also an author/historian. Her previous books include Montgomery County, A Pictorial History and Paul of Montgomery.

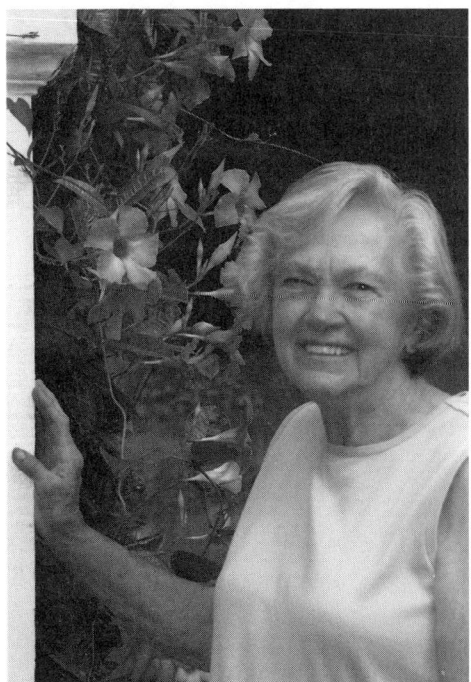